Let the waters bring forth abundantly.

GENESIS 1:20

Following pages:

A fish pond, Hawaii.

Chinese dip nets, Kerala, India.

A seaweed farm, Maisaka, Japan.

Harvesting shrimps, shrimp farm,
Oahu, Hawaii.

ELISABETH MANN BORGESE

with photographs by Robert Ketchum and others

SEAFARM

THE STORY OF AQUACULTURE

Harry N. Abrams, Inc., Publishers, New York

A portion of the text appeared originally in *The New Yorker.*

Library of Congress Cataloging in Publication Data

Borgese, Elisabeth Mann.
    Seafarm: the story of aquaculture.

    Continues the author's Drama of the Oceans.
    Bibliography: p.
    1. Aquaculture. I. Title.
SH135.B67   630'.9162   78-25680
ISBN 0-8109-1604-5

Library of Congress Catalogue Card Number: 78-25680

Printed and bound in Japan

CONTENTS

PART I

THE AGE OF AQUARIUS

PART II

MEADOWS AND FIELDS

PART III

FARM ANIMALS

PART IV

ON THE HORIZON

I
THE AGE OF
AQUARIUS

◀ A fish enclosure, Songkla, Thailand.

Laguna de Bay, the Philippines.

1

EXCURSION INTO THE PAST

We are approaching Iloilo airport...or Cochin...or Semarang...or Kwantung Province. Green fields give way to watery fields, flooded paddies that reflect the setting sun. The paddies merge into ponds whose earth embankments slope toward the bay. The bay itself is divided by bamboo fences into watery fields, with lanes for farm vehicles, paddled or motor-driven, leading to the floating huts of seafarmers and watchmen. Out to sea, rafts hold fish cages and rows of stakes are encrusted with black mussels.

We adjust quickly to thinking in terms of watery fields, to watching farm laborers till pond soil, fertilize water, herd fish. Eight or nine men, in water up to their hips, drag their net through the pond, and as the net closes in, the fish—carp, mullet, milkfish—begin to jump like silvery golden fireworks amidst the shouting men. The fish are measured and weighed and then returned to the pond. A proud owner produces his breeder carp as though he were trotting out his champion bull; he offers his largest prawn to honor the guest as though it were a fatted calf.

We soon accept that we are making our way through fish hatcheries rather than chicken hatcheries; meadows of drying seaweed rather than hay; rows of greenhouses containing eels rather than tomatoes.

We learn to rephrase questions about land use and productivity and labor and food in terms of water use. As we learn, the feeling grows within us that we are fortunate enough to witness something of enormous importance: one of the major transformations of the culture of man in his relationship to his environment.

The beginnings of agriculture, some time about 10,000 B.C., are shrouded in the darkness of prehistory. Agriculture probably originated in several places, among them certainly first the Middle East and then Central America. The advent of agriculture was not an "event," a "discovery," but a series of developments, an ongoing process, the emerging system overlapping and interacting with the previous system of hunting and gathering. The introduction of agriculture is one of the most important evolutionary changes. Man changed himself as he began to change his environment, turning forests and meadows into villages and fields, acculturing and transforming the beasts of the wilderness.

The introduction of aquaculture marks an evolutionary change of the same magnitude.

In some parts of the world fish farming and aquaculture go as far back, or almost as far back, as agriculture; in some parts the two systems have always been inseparably linked.

The ancient Greeks and Romans fattened fish in ponds; the Romans cultivated oysters. Tilapia may have been cultured in Egypt by the pharaohs. A tomb frieze of 2000 B.C. depicts the harvest of what appears to have been a tilapia pond culture. The fish that Saint Peter caught in the Sea of Galilee is said to have been tilapia, and Christ's miraculous feeding of the masses may symbolize the abundance of tilapia culture.

In China, aquaculture is rooted in antiquity. A treatise called *Fish Breeding,* the "Classic of Fish Culture," ascribed to Fan Li and dated 475 B.C., illustrates the spawning of captive carp and indicates that fish farming was widely practiced in China at that time. *Kwai Sin Chak Shik,* a book written during the Sung Dynasty in A.D. 1243, describes how carp fry were transported in bamboo baskets—much the way they are transported and traded today. The fry were collected in rivers and reared in ponds; this is recorded in *A Complete Book of Agriculture,* written in A.D. 1639.

The ancient Chinese understood that aquaculture is three-dimensional. Whereas a field is surface and topsoil on which the farmer can plant and harvest only one crop at a time, aquaculture occupies surface, water column, and bottom. Each dimension has its "ecological niches," which are home to different species of fauna and flora. Many fish prefer the sunlit surface of the water; there are also mid-water dwellers and bottom dwellers. Some aquatic creatures feed on phytoplankton, the lowest link in the food chain; others prefer zooplankton, the tiny animals, protozoa or brine shrimp; some are herbivores, others carnivores; still others feed on the wastes of other animals or plants. Relying on nature rather than on his own efforts, the aquaculturist can raise six or seven species of fish and crustaceans in a limited area. The Chinese are past masters of this method—polyculture—which they also apply to industry by designing systems of "factory linkages," and by integrating agriculture (always including aquaculture) and industry in their rural communes. Chinese polyculture, together with the Chinese carp, migrated from China to India, and, probably from there, in more recent centuries, to Malaysia, Formosa, Indonesia, and Thailand.

In ancient India, the kings and their scholars possessed an astonishing knowledge of the flora and fauna of their waters. About 600 B.C., Susruta gave a detailed account of the ecological classification of fish and of the ways in which their form and locomotion were adapted to their environment. Fish were even then being reared in ponds: Princess Khona, daughter-in-law of Varahamihira, recommended that vegetables be grown on the banks of the fishponds—an early example of the integration of agriculture and aquaculture.

King Asoka's Pillar Edict V (246 B.C.) spells out one of the earliest pieces of fish legislation. It prohibits the killing of certain undesirable or inedible species of fish—such as misshapen fish, or those that are snake-headed, or those that live only on flesh. Five kinds, in particular, are enumerated: shark, eel, the freshwater porpoise of the Ganges, skate or ray, and puff-fish. The edict also contains some fairly elaborate rules for the conservation of edible fish—pretty much the same range of food fish species as exist in the region today: the three "major" Indian carp (catla, rōhita, and mrigala), various catfish, and hilsa. These fish were declared inviolable and unsalable "on the three Cāturmāsīs, and on the Tishyā full moon, during three days, *viz.*, the fourteenth, fifteenth, and the first *tithi*, and invariably on every fast day. And during those same days no other classes of animals which are in the elephant park and the preserves of the fishermen must be killed." Modern Indian scholars marvel at the insight of these regulations, which protect the fish during the height of their spawning seasons.

The ancient Indians kept fish in tanks, lakes, and rivers, and intensified and protected their cultivation during periods of hunger. King Somesvara III, who reigned from A.D. 1126 to 1138, composed an encyclopedic Sanskrit poem which deals, among other things, with the culture of fish for sport. The poem prescribed with astonishing scientific modernity how the fish are to be fed and fattened:

> Now about feeding the fish. One should feed them both in the evening and in the morning. Their food consists of the following things: cakes and balls of sesame seeds, powdered or dried rice, and flour of dried chick-peas mixed with cooked rice.
>
> Make balls of the size of a crabapple and feed kahlava and the like fish on them.
>
> Powdered with oily substances, the stuff should be thrown in the water, and the water then stirred.
>
> Make balls the size of a plum from the sesame cake and seed preparation mixed with cooked rice, and feed rōhita and the like fish on them.
>
> With the powder of dried safflower and flour of dried barley both

mixed with cooked rice: Make balls of the size of a mango and feed the badisá fish on them.

Take leaves of the crabapple tree, crush them, and mix them with dried barley.

Make balls as big as an āmla fruit (a kind of gooseberry), and let the expert scatter them to feed the fish.

The expert should feed the kauvāka fish on bits of lung and hind muscle. One should feed the pathīna fish on bits of foul-smelling meat. The expert should feed the simhatunda fish on dhichakas.

The fisherman should take care that he feeds the maralī fish on the flesh of crabs.

The expert should feed turtles on the roasted flesh of mice.

One should feed the small fish on earthworms. On the steps of the banks from where the water is taken.

One should take care that one feeds the fish in the manner detailed above.

The royal attendants, having thus fed the fishes, should go and inform the king.

The variety and complexity of these recipes is surprising. King Somesvara obviously knew about the herbivorous, carnivorous, or detritus-feeding habits of different species, the needs of small fry, and the usefulness of recycling various kinds of waste and offal.

The pond system of Java is an achievement of human labor as remarkable as the building of the pyramids. In Java there are some 200,000 acres of ponds. Since approximately 2,500 cubic yards of soil have to be removed to construct one acre of pond, a total of 500 million cubic yards of soil had to be removed by hand to give the ponds their present shape. The Javanese law code *Kutara Menawa,* written probably around A.D. 1400, provides the earliest information. It seems that the first of these ponds were built early in the second millennium. The code prescribes punitive measures "against those who steal fish from a freshwater pond (*siwakan*) or a saltwater pond (*tambak*)." Legend has it that the very first ones were built between A.D. 1200 and 1400 by convicts sent by the Hindu rulers to the salt marshes. Barred from agriculture and from shipping—even from wearing clothes—these convicts dammed up

Farmer Uymura's loko kuapa, Oahu, Hawaii.

Farmer Uymura manages his farm with the help of his wife.

creeks or bights to catch fish and prawns to survive. The pond culture expanded from eastern Java to the center and west of the island.

The ancient Polynesians also had systems of aquaculture. When they settled in Hawaii, about A.D. 1000, they brought their concepts and techniques of pond construction and management with them. They studded the islands with ponds—ponds near the shore, in bays, skillfully dammed by walls of basalt and coral; ponds inshore, dug and bunded. There were about 210 of them on Hawaii, Maui, Molokai, Oahu, and Kauai, where mullet and milkfish were raised by king and commoner. S. M. Kamakau, a nineteenth-century Hawaiian chronicler, observed that "Fishponds were things that beautified the land, and a land with many fishponds was called 'fat.' " Most of the ponds are no longer in use, and their ancient walls are now in ruins. Others, however, are amazingly well preserved and under active management. Today in Hawaii there is a great movement to restore the ponds and to reactivate the ancient industry.

Farmer Uymura's *loko kuapa* pond was built in the sixteenth century. The ancient *kuapa,* or volcanic rock wall, needed little repair. The sluice gates, or *makaha,* were in place, and only the wooden parts had to be renovated. Every day, using his small motorized craft, Uymura catches his mullet fry beyond the wall, in the near-shore waters of the sea. He nurses them for a couple of weeks in a series of old bath tubs acquired at the junk store. Then he stocks them in his pond. The pond is fertile, a rich algae green, and no additional feeding is needed. At harvest time, after about four months, the mullets are netted by the sluice gate. Farmer Uymura manages his pond with the help of his wife. They have made a good living and built themselves a fine home. But times are not getting easier. More capital is needed to modernize production methods and make fish prices competitive than is available to a traditional fish farmer and his wife. And Uymura is seriously thinking about selling his paradise.

2

SEVEN REASONS FOR AQUACULTURE

Aquaculture remained stationary through the millennia like the cultures in which it was embedded. Agriculture, however, kept pace with the scientific and technological revolutions of the Western world. It became industry; it became "agribusiness." It kept up with the increasing demands of multiplying populations, while food from the seas and lakes and ponds and rivers contributed only a minor and diminishing share. More than 96 percent of the world's food comes from the land. It is not surprising, therefore, that economists and planners, in their attempts to banish the specter of mass hunger, have tended to overlook the potential contribution of aquaculture and seafarming.

Yet there are a number of global factors which today command the arrival of aquaculture, just as analogous factors commanded the rise of agriculture over the past ten millennia.

1. Certainly in the developing world there is still room for improvement and expansion of agriculture: cropping can be intensified; irrigation systems can be perfected. Some fallow land can be brought under cultivation. Forests, especially tropical rain forests, can be made into fields, although the long-term ecological consequences may be disastrous. Yet even if all possible improvements are made in production, storage, pest control, fertilizing, fuel, transportation and, above all, distribution, we will still need more food—an estimated increase of at least 3.5 percent a year. How long can we continue to accept the fact that more than half—perhaps two-thirds—of the world's people suffer chronic undernourishment, their minds and bodies stunted by protein deficiencies?

We have to recognize that the possibilities of expanding agriculture are very limited, and that the limits themselves are shrinking. In simple terms: the more people there are, the more food is needed; yet the more people there are, the less land is available for food production.

The possibilities of expanding aquaculture, however, are for all practical purposes unlimited.

In 1970, estimated total production through aquaculture amounted to 2.6 million tons; by 1976, it had risen to over 6 million tons, of which 66 percent consisted of freshwater, brackish-water, and marine fish, about

16.2 percent of mollusks, 17.5 percent of seaweeds, and 0.3 percent of crustaceans. The most conservative estimates—based on existing simple technologies and systems of culture—project a doubling of world production in ten years, and at least a fivefold increase by the end of this century. According to more optimistic forecasts, the increase in this period may be as much as tenfold.

Asia, which accounts for 85 percent of the world's aquaculture production, presently utilizes just over 2 million hectares for this purpose (a hectare equals about two and one-half acres). It has been estimated that *over 20 million hectares more* are available for development in that area alone. Such an expansion would not conflict with the expansion of human habitat, since the soil is either too saline for agricultural use, or not stable enough for construction. Conflicts might arise between natural ecology and the development of aquaculture. In the estuaries and mangrove swamps that are the natural spawning and rearing grounds for many fish of the sea, great care must be taken not to upset the ecology on which natural marine life depends. But aquaculture can be planned in such a way that it not only conserves but even enhances the natural production of the open sea. Beyond that, there is tremendous scope for aquaculture in bays, coves, and inland seas where fish and prawn and squid can be farmed in walled-off or fenced fields or in floating cages, and mussels and oysters can be grown on rafts and stakes.

Guimaras Island is a small tropical paradise an hour's brisk ride in a motored outrigger canoe from the South East Asia Fisheries Development Center (SEAFDEC) on Iloilo in the Philippines. There is a fishing village on Guimaras, but natural fisheries are not abundant. The island has nine beautiful natural coves, worn out of the volcanic rock. The water above the coral reef is beautifully clear.

A seafarming project is now in progress on this island which might be adapted to many islands with a similar geological conformation, providing labor and nourishment to the inhabitants.

The people of Guimaras are walling off the coves with volcanic rock and coral. It is hard work. To wall off just one cove has taken eight months. The cove, thus converted into a farm, has been stocked with marine milkfish so tame they will eat from your hand. The other coves, when converted, will be stocked with catfish, prawn, and squid; two of them will support polycultures.

The total cost of construction of this seafarm is estimated as one million pesos (U.S. $140,000). The annual income from production is estimated as half of this amount!

Beyond bays, coves, and inland seas, the open ocean offers unlimited space to free-ranching sea "cattle" reared in land-based hatcheries and nurseries. Tuna and salmon, for example, can be reared to the stage where they can successfully cope with predators or with the inclemencies of nature; their chance of survival thus increases a hundredfold. They can be "transplanted" to where the pasture is greenest. Salmon from the North American and Japanese shores can be moved to the Soviet Arctic; Antarctic krill, overabundant since the depletion of the great whales, would nourish vast numbers of salmon, whose route of migration could be modified by selective breeding and postnatal "imprinting."

2. Agriculture is bidimensional. Aquaculture, as the ancient Chinese knew, is three-dimensional, offering multiple crops and savings in material and labor costs.

Farmer Maharajganj's polyculture, on the road from Calcutta to the Bay of Bengal, is a flourishing enterprise. The area, covering 4.6 hectares, is divided into 16 ponds. The three major Indian carp are raised together with three species of Chinese carp, utilizing every niche of the ecosystem. Artificial food, in the form of groundnut oil cake and rice bran, in equal parts, is added, totaling between 2 and 3 percent of the fishes' bodyweight. Farmer Maharajganj has thirteen workers employed full time. The cost of production is about 2.5 rupees per kilogram of fish. Since the sales price is four times as much, there is a handsome margin of profit.

The world ocean is one vast polyculture. To learn to understand and wisely manage the interaction of its multiple uses is the almost superhuman task of the new science of ocean management.

3. Agriculture utilizes long food chains to provide protein to people. The primary product is starchy staples. Aquaculture yields protein crops, and aquatic organisms are often more efficient converters of primary foods than are land animals. Dr. C.P. Idyll, Chairman of the Division of Fishery Sciences of the Institute of Marine and Atmospheric Sciences at the University of Miami, Florida, claims that the content of many kinds of fish protein is equal to that of protein from beef, veal, lamb, and pork, the relative amounts of the essential amino acids being almost identical. At the same time, the assimilability of fish protein is higher than that of other animal protein. Furthermore, the cost of producing animal protein from fish is significantly lower than the cost from land animals. On a worldwide basis it costs half as much to produce a ton of protein from fish as from beef, a third as much as from pork. In labor expended, fish requires a third as many man-hours to process as pork.

4. We do not know what will happen to the world's climate during the

23

final years of our century. Meteorologists are divided in their opinions. Many of them believe that the postglacial period of the last ten thousand years which has given rise to our present civilizations is coming to an end and that the world's climate is cooling. An expanding snow cover may reflect an increasing proportion of sun heat back into space and thus decrease temperatures, and this, in turn, will further expand the snow cover. The process may be accelerating. On the other hand, there are meteorologists who hold that the planet may be warming up, due to an interaction of natural and man-made causes: among the latter, the emission of too much carbon dioxide from our heedlessly energy-intensive society. All meteorologists agree, however, that no matter in which direction the change may go, the transitional period will be characterized by climatic instability, excessive heat alternating with excessive cold, accompanied by droughts, floods, other natural catastrophes, and, regionally, by a shortening of the agricultural growing season.

Paradoxically, aquaculture is less severely and less immediately affected by the weather than is agriculture. In times of drought the level of ponds may drop but still suffice for aquaculture when irrigation systems collapse; brackish water may be abundant but unusable for agriculture; ocean spaces may remain unaffected or be more abundantly fertilized by upwellings caused by the cooling of surface waters, while agriculture languishes or dies altogether. The statistics of the 1972 drought in Southeast Asia support this assumption.

As the receding glaciers made way for agriculture ten thousand years ago, so the climatic change of our own era may spur the expansion of aquaculture. A number of minor man-made circumstances appear to reinforce this natural evolutionary trend.

5. Hunting is not a way of life that is suited to industrialization. Industrialized hunting is a self-contradiction. Industrialized hunting destroys itself. More and more sophisticated gear is being applied to capturing a natural resource that is becoming scarcer and scarcer, not only because of ruthless overfishing but also because of industrial pollution and interference with natural breeding grounds. Capture fishery is reaching its outer limits. One fishing industry after another is collapsing, among them the whaling industry and the anchoveta industry. The whole Northern Atlantic is "overfished," and estimates of the "maximum" or "optimum" sustainable yield of commercial fisheries have had to be drastically reduced. With wise, internationally controlled management, present catches—actually on the downslope of the curve—can perhaps be stabilized once more. The curve can perhaps be turned around: from

24

Bamboo-fenced lane, milkfish farms, Laguna de Bay, the Philippines.

A fish-and-shrimp enclosure; and a rock-enclosed cove, Guimaras, the Philippines.

Aquaculture ponds near Semarang, Indonesia.

Capture, a phase of culture: ▶
fishermen harvesting stake nets
in the Bay of Cochin, Kerala,
India.

Vizhingam, a
traditional fishing
village, Madras,
India.

Vizhingam, the Catholic church.

Fisherman hauling in
the night's catch,
Vizhingam.

A brew of cow dung
and tamarind roots,
Vizhingam.

Overleaf:

Culturing mussels and
lobsters, Vizhingam.

Birds and boats, Vizhingam.

50 to 60 million tons worldwide we may perhaps reach 100 million; but that is the outer limit, and it is not all that much. If science and technology are to be applied to the production of aquatic food and fiber, the whole system must be changed. It must be *productive* rather than *extractive,* integrated into *farming* rather than *hunting*.

Most—about 85 to 90 percent—commercial fish have been extracted in the nutrient-rich waters above the continental shelf, in an area less than 200 miles from the coasts.

The coastal waters of the Northern hemisphere, bordered by rich, industrialized states with powerful and sophisticated fishing fleets, were the first to show symptoms of resource exhaustion. As this happened, their fleets set out for distant shores, to deplete the coastal waters of developing countries unable to explore and exploit the resources.

6. Then came the energy crisis. Soaring fuel prices threatened to drive distant-water fishery out of business. The natural resource, scarce off the shores of the homeland, was becoming too expensive in the distant seas.

7. As if that were not enough, the new Law of the Sea, extending coastal-state sovereignty over all natural resources within a 200-mile "exclusive economic zone," effectively did away with the time-honored "freedom to fish" on which the capture economy had been based. It is a foregone conclusion that the adoption of the exclusive economic zone concept, adding legal restrictions to the already active economic restrictions on distant-water fishing, will contribute to a decline in the world fish catch.

For coastal states will not be able, in many cases, to replace the departing distant-water fisheries. The poorer ones among them do not even have the research ships to explore the natural resources of their area. And that, again, is a reason why they may turn to aquaculture and seafarming. Why go through the costly and self-defeating phase of industrialized hunting when they can move smoothly to farming?

Thus the array of factors advancing the arrival of aquaculture and seafarming is quite impressive. It is not surprising that experts such as John Bardach (marine biologist at the University of Hawaii) and others, predict that fisheries will be profoundly transformed over the next fifty years and that all fish will be cultured in some way or, to put it more precisely, that there will be human intervention in one or more stages of the life cycles of all living aquatic resources.

A change of this magnitude pushes back the limits to growth. For all practical purposes, these limits are in man, in his capacity to organize, utilize, and distribute—not in the carrying capacity of the planet.

3

A NET OF CONNECTED MESHES

Vizhingam is a fishing village on the coast of India near Madras. It looks today as it may have looked for a thousand years. The few hundred huts, built of mud and straw, house as many as twenty thousand men, women, and children on the narrow strip of beach. In front of the houses are the fishing boats, partly anchored in the water, partly pulled up on the beach—dug-out canoes with red sails, catamarans with outboard motors. There are almost as many boats as there are huts. All of the boats, like the huts, have roofs of palm leaves.

It is morning and Vizhingam is busy. Between the huts, nets are being made and mended by old men and women, the part of the population that is not seagoing. A row of large cans is set over a fire of bamboo straw, and in the cans a reddish-brown liquid is bubbling and foaming. The liquid is cow dung and tamarind roots. In this brew the netmakers dip their nets to strengthen them against corrosion by the saltwater. The brew, with its high ammonia content, also acts as an antiseptic.

About half of the population is crowded near the water. The night's catch is being carried from the boats in tubs and baskets. The crows, almost as numerous as the people, cruise over the village, alighting on boats and tubs to snatch their share of the bounty.

When the catch is good, it may yield as much as 1,500 rupees, which, divided among fifteen men, compares favorably with the workingman's wage in Delhi of 5 rupees a day, in Kerala of 12 to 15. A part of this, however, must be paid to the Church.

On the left, towering over the village, is a Christian church; on the right a mosque. The half of the village in the shadows of the church is Christian, the other half Moslem. Sometimes tension between the two halves erupts in fierce fighting with knives and staves.

Not far from the mosque is a hut, just like the others, its walls made of woven coconut palm, its roof of rice stubble. This is the Lobster and Oyster Culture Farm, Kovalan, a branch of the Central Marine Fisheries Institute.

Inside is an array of plastic tubs and plastic-lined tanks, pictures and charts, and a table and benches. Unlike the other huts, this one is equipped with electricity which powers the aeration system to tubs and tanks. And there are some cages holding oyster trays.

◄ Fish farmers with their catch, near Calcutta, India.

The oysters are there for the surgical implant of the artificial pearl nucleus. Mastered by the Japanese, this delicate operation is now standard practice in India. After the operation, the oysters convalesce in culture cages of heavy wire frames in tubs in the hut. Then they are taken out on a catamaran to the raft.

There are 8,000 oysters on one raft. There are six rafts, made of casuarina wood. The casuarina is fast-growing: in one year a forest of trees grows all along the beach.

From the rafts hang ropes of coconut fiber with tiles attached to them. Mussel spat (young) cling to the tiles. They are collected when they are about one centimeter long, and fastened to other ropes in canvas stockings. As they grow they attach themselves firmly to the ropes and burst the canvas. They grow very rapidly: 12 millimeters per month. In eight months they grow to marketable size. Total production is 5 to 6 tons per raft in eight months. Production per raft could be increased fivefold, to reach the productivity of Spain or Holland; there, however, the growing time required is four times as long.

Lobster larvae which, in the wild, attach themselves to rocks, are attracted to the tiles. From February to May, thirty to thirty-five lobster larvae daily congregate on each of thirty tiles. They are transferred to the tubs and tanks in the hut, where they feed on mussels and grow. In only two years—one inside the hut, one in cages in the sea—they reach marketable size. They grow in spurts, while molting. Research is under way to make them molt more often and, perhaps, grow twice as fast.

Apart from the research, the labor required for each raft is only two men working about two hours a day. It is a part-time job for the fishermen of Vizhingam. They quickly learn the skills of aquaculture. During their working day, they move between magic and religion and science and technology.

Capture and Culture

Capture and culture are linked in a number of ways.

Fry Production
Many farmed fishes do not breed in captivity, and although there is a growing science of "induced spawning" and hatching and nursing fry and fingerlings, the multibillion fry market is still largely supplied by fry caught in the wild. Carp and catfish, mullet, puntius and milkfish and eel are caught in the wild. Methods differ from country to country and from region to region. A vast array of traps and nets, basins, pails, and cups; earthen jars, pandanus bags, lamps, pouches, and purses have evolved

44

over the centuries. Fry gathering is seasonal. Since it occupies only about three months out of the year, fry gatherers need other work. Many of them are fishermen, others farmers or laborers.

The fry market is a very special kind of market.

In Sukabumi, Indonesia, there is a fry market every morning. A well-designed facility, with sun roof, troughs, and freshwater flushings is provided by the government. The traders come from near and far, on foot, on bicycles or motorized vehicles, each carrying two black baskets balanced on a shoulder-borne staff. The baskets are made of palm straw and are tarred to make them waterproof. Each basket contains two or three inches of water filled with young fry of all sorts: tilapia and carp and Sepat siam and Tawes and Nilem and catfish and the kissing gourami. (Lest anyone should wonder about the name of the "kissing gourami," the fact is that he kisses. He kisses anybody and anything in reach, with pursed thick lips. Fish psychologists explain the kissing as a gesture of threat.)

The dealers set their wares down on the ground, stirring the water with their hands to aerate it. There may be 700 baskets set down in one morning. As the sun grows hotter, the traders either place their baskets in the cooling troughs, or change the water. They smoke and talk and praise their wares and bargain. They count their tiny fish, swiftly and delicately passing them from one measuring cup to another, chanting the numbers as though they were prayers.

The buyers come from near and far. They, too, come with their baskets or, if the journey is long, with styrofoam boxes holding plastic bags filled with water and compressed oxygen. This way fry can travel long distances unharmed.

By noon most of the fry has changed hands, and the market closes. Sellers go home for more fry; buyers stock their acquisitions in nursing and rearing ponds: about 5,000 fry per hectare. They may raise them for a whole season to "market size" or they may keep them for a four-week period—as long as there is a pond or rice paddy available between harvesting and planting. Their purchase may put on 150 kilos during that period, and each fish may have grown 12 centimeters longer. Then they may take these fingerlings back to the market and net a handsome profit of $200 per hectare. With fish changing hands so many times at different stages of their life cycle, it is often difficult to assess the total value of fish culture in a given region.

A part of this market is supplied from the wild by fry catchers. However, there is an increase in the number of fry *growers*—farmers who proudly maintain carefully selected breeders in their ponds. One farmer, just near the fry market of Sukabumi, keeps his breeders in a little

A farmer keeps his breeders in the middle of his living room, Sukabumi, Indonesia.

indoor pool right in the middle of his home. The man is both a carpenter and a farmer, so the little indoor pool is used to wash his working wood, to wash his rice before planting, as well as his pots and pans. The breeders seem well adjusted to the many uses of their small environment. They are sociable, communicative, and curious.

A few carp, tilapia, catfish, and others spawn naturally in small hatching ponds sometimes not much larger than a bathtub. All the farmer has to do is to collect the glistening spawn during the cool hours of the morning from the *kakabans* woven from the horsehair-like fibers of the Indjuk palm which are placed in the pond for this purpose. One farmer may gather as many as 40 cups a day. Each cup contains 4,000 spawn. He transfers them to his well-cleaned and fertilized nursing pond, whence they go to the fry market.

Other fry, billions of them—fry of fish which do not spawn spontaneously in captivity—come from government hatcheries where selected breeders are given injections of hormones. Spawning is thus artificially induced. The hormones are extracted from the pituitary glands of fish, mostly of the same species as the breeder to be injected. Sometimes another species, mostly common carp, is used. In some instances hormones from mammals, including man, are effective.

The method was developed in the 1930s by Brazilian biologists and signified a major breakthrough in fish farming. The extracting of the pituitary gland from the donor fish is quite simple. The fish can be "borrowed" from the fish market—even if it has been on ice for several hours, the pituitary is still effective—and the gland, which is located below the brain on the floor of the brain box, is removed, using a butcher knife, a needle, and forceps. The fish, minus pituitary gland, can still be sold for food. Even in a crowded and noisy fish market an experienced worker can collect an average of fifty to sixty glands in the space of one hour. The extracted glands are immediately put in alcohol, which dehydrates and defats them. Then they must be stored in a refrigerator. Before use they are weighed, ground up in a tissue homogenizer, mixed with distilled water and glycerine, and poured into vials.

The breeder fish are kept inside a hand net, or, like racehorses, they have a kind of sleeve or hood slipped over their heads. The injections are given intramuscularly near the tail region, or intraperitoneally (body cavity injections).

Injected fish may spawn spontaneously, in specially prepared nuptial chambers—in India they are called *hapas*—where spawning usually takes place in three to six hours after a second injection. Alternatively, both males and females may be "stripped" of their milt and eggs, which are then mixed in a container, either "dry" or "wet," that is, with the addition of water. The fertilized eggs are then placed in hatching jars provided with running water and oxygen.

Artificially induced spawning, by various methods, has become a routine activity for many species of fish and crustaceans. Near Semarang, in Java, for instance, the Indonesian government is running an impressive demonstration hatchery where the farmers of the region are trained in fry production, and the hatchery's own production is quite substantial. Nursing and rearing ponds occupy 4½ hectares (a little over 11 acres). They are drained, sun-dried, and fertilized with cow dung once a month. Methods and materials are simple, and easily taught. The farm contains facilities, including dormitories, for twelve trainees who need stay for only one or two weeks. Eighty million fry a year are produced by this demonstration farm alone, 95 percent of them tawes and Mata Merah, but also some tilapia and kissing gourami.

The cost of running the demonstration farm amounts to 400,000 rupias. The income is threefold: 1,200,000 rupias. The director of the farm, Mr. Taib, is a man of only elementary education, and he knows what farmers need.

For other fish species, induced spawning is still in the experimental stage. First, the cause of the failure to spawn in captivity has to be ascertained. Shrimp, for example, have something in the chemistry of their eye stalk that inhibits sexual maturation in captivity. Now, as the Bible says: "If thine eye offend thee, pluck it out" (Matthew 18:9). It has been discovered that ablation of the eye stalk removes this chemical inhibitor, and maturation of the gonads and spawning follow within a couple of weeks. The ablation of only one eye stalk is sufficient and does not inflict much suffering. Bilateral ablation, blinding the prospective breeder, causes too grave a trauma and results in death.

Spawning in shrimps can also be induced by manipulating the environment, in some cases by raising water temperature or changing water salinity, in others by raising the water level or increasing the current. One shrimp hatchery in India, where spawning was induced by such environmental management, got a surprising result—a generation of one-eyed shrimps—hundreds of thousands of them. A strange dialogue between nature and culture, as though based on computer logic, with an inversion of cause and effect.

Milkfish, which are a rich source of animal protein for the people of the Philippine Islands, the Indonesian archipelago, and Taiwan Province, are notorious for their failure to breed in captivity.

The milkfish in Laguna de Bay in the Philippines should not even know they are captives. Yet somehow they do. A fatty substance forms round their reproductive organs, which atrophy. Although milkfish are

A crowd of ▶ fingerlings, a fry market, Sukabumi, Indonesia.

A fry counter in Hong Kong. He buys across the river in China.

A spawning pond; and a proud owner showing his breeder carp, Sukabumi, Indonesia.

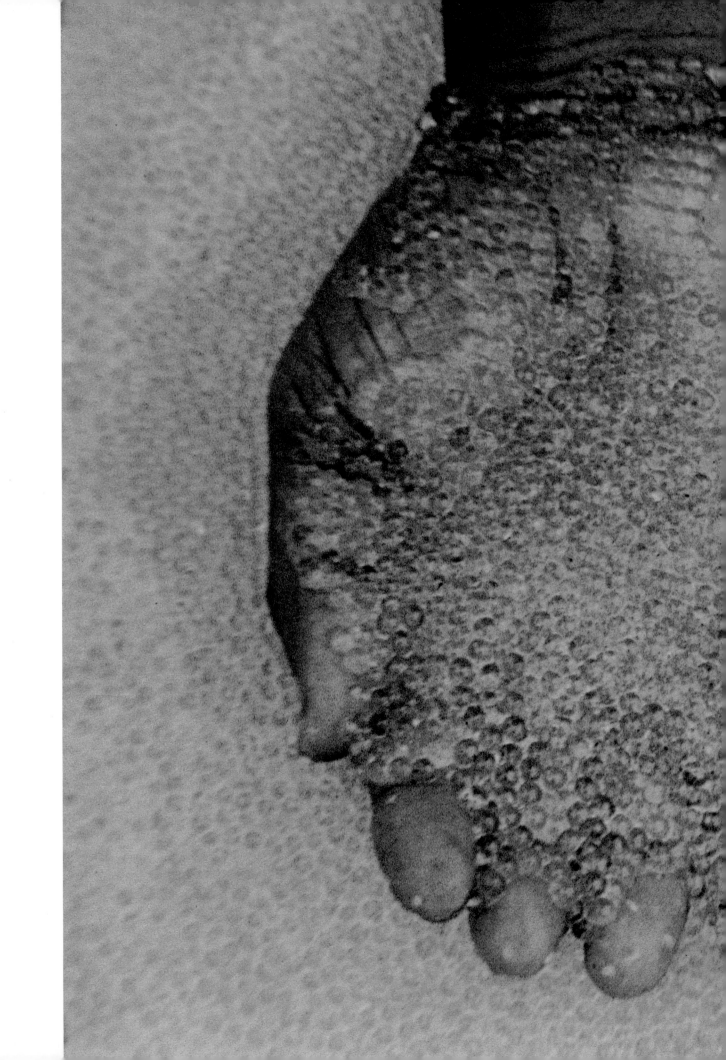

Fertilized eggs.

Fresh spawn, Sukabumi, Indonesia.

Fish food: Rice bran and fish meal, with added minerals, vitamins and cheese. Inland Fisheries Institute and Demonstration Farm, Bogor, Indonesia.

Inland Fisheries Institute and Demonstration Farm, Bogor, Indonesia. The hatchery. In the rear basin, water is filtered through rocks.

Ponds and feeder canal, the demonstration hatchery, near Semarang, Indonesia.

◄ Feeding fish in Japanese-type running water
pond where fish can be stocked at great density.
Inland Fisheries Institute and Demonstration
Farm, Bogor, Indonesia.

Sluice gates, the demonstration hatchery, near Semarang, Indonesia.

cultured in vast quantities—this region annually produces about 150,000 metric tons—the quantity is limited by the number of fry which can be caught by the fry catcher in a given year. When it becomes possible to breed this beautiful fish in captivity, the supply of its tasty, protein-rich flesh will be plentiful and consistent. A breakthrough was recently made at one of the SEAFDEC stations in the Philippines. A six-year-old, sixteen-pound female was induced to mature her eggs—hundreds of thousands of them—and a male was rushed to the scene by plane in his oxygenated plastic bag, to yield his fertilizing milt. The fry are said to be doing fine.

Fry catching thus constitutes a link between capture and culture fisheries. As research and development in artificially induced spawning progress, this link is bound to disappear, causing some displacement

Fry is caught in a dip net for inspection, near Semarang, Indonesia.

within the displacement. The fry catcher will have to be retrained for other roles within the greater shift from capture to culture.

Sea Ranching

While one link is weakening, another is getting stronger—namely, the release of artificially hatched and nursed young fish into the marine environment. This, again, can take a variety of forms, some of them already "commercialized," some as yet in an experimental stage. Anadromous fish, that is, fish that breed in rivers, lakes, or streams, from which the young return to the sea (for example, salmon), are obvious candidates for this kind of "sea ranching." But so are intraocean migrants, such as tuna, which can be bred and reared on land and then released to reinforce commercially depleted stocks. Nature is appallingly wasteful; at least as wasteful as man—who, after all, is part of nature. Of a thousand fertilized fish eggs perhaps one, in nature, may attain maturity and reproduction. Man can, to a large extent, redress this situation by nursing fish brood in a controlled situation beyond the critical point at which the young fish can successfully fend for themselves.

Wasteful nature also leaves ecological "niches" in the ocean polyculture unused. Here, too, man may come to the rescue. Russian scientists have taken fish eggs, larvae or fry, of species not indigenous to the waters off their shores and transplanted them successfully: mullet from the Black Sea to the Caspian; striped bass from the United States to the Black Sea, the Sea of Azov, and reservoirs; steelhead from the United States and chum salmon from Japan to the White Sea. The Russians have made quite a science of acclimating fish from one part of the world to another. There are a number of factors that determine the success or failure of these ventures. A first set of problems concerns the fish themselves, individually and collectively. Will the immigrants be able to adapt to an environment where water salinity, temperature, and circulation patterns may be somewhat different from their home grounds? Where the days may be shorter or longer, the food supply taste different, and unfamiliar predators lurk? If enough of the immigrants make it—sometimes aided by transitional stages during which home water is mixed with increasing amounts of host water in nursery containers—will selective breeding take place and increase the survival rates of future generations?

Hybridizing with the natives of the host waters may facilitate acclimating. This development may be spontaneous, or it may be scientifically planned. Soviet scientists, again, are masters in this science. They have accomplished astonishing feats—for example, in hybridizing different species of sturgeon. The famous caviar-bearing Beluga sturgeon of the Caspian, a species dwindling alarmingly because of pollution and overfishing, has now been successfully bred to the much smaller freshwater-dwelling sterlet. The offspring of this marriage has been named "Bester," after both parents. It is an enormous, Beluga-like freshwater dweller, so large that the female carries as much as 50 kilos of caviar during one pregnancy! To enhance the survival rate of this prodigious animal the Russians have perfected a surgical technique in which she is relieved of her caviar, stitched up, and released after ten minutes or so, unharmed, into the water—a speedy and efficient little fish Caesarean which a female may undergo as many as five times during her sexually mature life.

Another set of problems concerns the host environment. The immigrants may carry diseases which, while endemic and relatively harmless at home, are liable to become epidemic and assume catastrophic proportions in the alien and unprepared host environment. They may be infested with parasites that attack the host fauna and flora; or the seed of extraneous species, not planned to be included, may slip aboard, illegal immigrants which may create undreamed-of problems.

Then again, the legal immigrants, the planned transplantees, may be so successful that they burst the invisible walls of their planned ecological niche and take over the whole host environment, preying upon and depleting indigenous fauna and flora, starving and crowding out competing domestic marine life. Such shake-ups of the local ecological balance have taken place in nature endless times, accountably and unaccountably.

Such problems must be studied case by case. Owing to the complex interaction between natural and man-made circumstances, they can never be solved completely. And while the intended results of the acculturation may become clear within two or three years, the side effects on the environment may reveal themselves only several generations later. It is estimated that over two billion juvenile anadromous fish annually are artificially produced and released into the fresh and marine waters of the world, mainly from government hatcheries. Some experts, especially in the United States and England, have expressed doubts as to the commercial viability of this activity, but other countries have taken it up on a large scale. Thus the Baltic, an endangered sea which seemed doomed only a decade ago, has been stocked with salmon, trout, and sturgeon which are fed in stations near the coast and netted by prosperous seafarmers. T. V. R. Pillay of the Food and Agriculture Organization of the United Nations (FAO) reports that according to recent studies in North America, every dollar spent for hatchery rearing and releasing coho salmon smolts has returned seven dollars through capture fishery. Large-scale release of hatchery-reared shrimp in the Inland Sea of Japan is reported to have caused substantial improvement in the local shrimp fishery, with a cost-benefit ratio occasionally as high as 1 to 10.

A nature-loving French couple with their charming, Tamil-speaking little daughter have established a green turtle farm just behind the beach near Madras. For a modest premium paid by the government the local population, especially the children, collect thousands of wild turtle eggs on the beach. They bring them to the turtle farm, where they are buried in the sand by the diligent French family. The turtle farm looks like a small cemetery. Each little grave, holding a few hundred eggs, is marked with a little flag. Protected from marauding dogs and other predators, the eggs hatch in the warm sand. During the cool hours of the morning, the

Baby turtles make their way to the sea, near Madras, India.

French family probes the graves, digs up and counts the hatchling turtles. In Tamil. In French. In very French-Tamil English. Thousands every day. They release them near the water's edge and supervise their course, first hesitant, then joyous, into the waves. When the last little turtle has been engulfed by the surf they turn away, as do the ever-present crows frustrated in their hopes of making a meal of the hatchling turtles.

Human protection ceases at the water's edge. How much this laudable activity actually contributes to replenishing the dwindling stocks of wild green turtles is not known.

Bait Production
There is yet a third interaction, or link, between capture and culture fisheries, and that is the culturing of baitfish for sport and commercial capture fisheries.

Fishing activities are often limited by the lack of suitable bait, and much time and energy are spent on catching baitfish to catch the real thing. Skipjack tuna are still quite abundant in the tropical Pacific, but fishing is limited by the scarcity of bait. This bait must have certain

A green sea turtle.

A woman planting rice
in a rice and fish pond,
Thailand.

Netting carp, tilapia, and catfish on a fish farm integrated with a sewage purification plant, near Calcutta, India.

Processing shrimp on a fish and shrimp farm,
Cochin, Kerala, India.

A prawn and shrimp cannery, Cochin, Kerala, India.

qualities: among other things, it must be of a certain size; it must be silvery to attract the fish; it must be able to survive for extended periods aboard a fishing vessel; and it must be available throughout the year.

With the rising cost of fuel it has become increasingly uneconomical to go to sea to catch anchovies, sardines, or young herrings as bait. Therefore, attempts have been made in recent years to culture bait in high-density hatcheries. The species that have turned out most promisingly are two kinds of topminnow, but a good many other species have also proved successful. In Hawaii they spawn in an array of simple troughs. After hatching, the eggs are placed in nurseries, fed on a variety of natural and artificial feed, such as fishmeal and chicken offal, and grow to baitfish size within three months. The survival rate is exceptionally high, and the cost of producing them is substantially lower than the cost of capturing live bait.

Capture fisheries will not disappear, simply to be replaced by cultures. Capture and culture will continue to interact in different ways. Culture is being inserted into capture. Capture is becoming, increasingly, merely a phase of culture.

Aquaculture and Agriculture

Aquaculture interacts in a variety of ways with agriculture. In simple, natural ways, as part of cycling and recycling systems, so that nothing is lost. Everything is used, and used again.

Fish and Rice
A paddy farmer floods his paddies when the tides are high and the monsoon swells rivers, lakes, and reservoirs. The tidal water, entering through simple wooden sluices, carries young prawns from the sea, and mullet, pearlspot, and milkfish. They thrive on algal growth and the decomposing rice stalks, which provide shelter as well as food. After three months the sluices are opened again, and fish and prawns are caught in nets attached to the sluice gates—as much as 350 kilos of fish per acre of paddy field. A simple operation that requires neither much effort nor much money. The fish and prawns fetch a price six times higher than the rice.

Fish and Ducks
When the fish have been harvested, the farmer brings in a flock of ducks—hundreds of them. They fatten on the fish left behind when the paddy was drained. They also fertilize the soil with their droppings, thus

Fish and duck farm, Hong Kong.

increasing the next harvest. There is still time to get in a harvest of vegetables before the next monsoon sets in, cleansing and flooding the paddy. Then the rice is planted again, a special strain, hardy and resistant to the salinity of this brackish water.

In Hong Kong it has been calculated that duck droppings in a fish pond can increase the pond's fertility and reduce the need for additional feeding by a quarter. The ducks, which live in an enclosure just above the pond, are fed small fish, sorghum, broken rice, and other food grains. And including ducks in the culture system about doubles its protein production.

Fish and Coconut

The paddies are bounded with earth embankments. These bunds are reinforced by the sprawling roots of the ever-present coconut palm. Tender coconut, the top chopped off with a jungle knife and a straw inserted in its abundant milk, is a symbol of hospitality—clean, refreshing, energizing. Coconut milk is also the base for various culinary delights. Fish curries may be cooked in coconut milk, or bananas—for a simple, delicate, cooling dessert.

The fiber of the coconut's husk is the basis of an extensive industry of coir rope, and mats and nets. Coconut palm branches are used as spawn collectors where aquaculture becomes more sophisticated. They are also stuck into ponds, here and there, to prevent poachers from stealthily dragging their nets by night. Palm leaves and straw provide walls and roofs for the well-aerated and insulated huts in which the farming community lives. Thus, a one-acre farm, integrating aquaculture and agriculture, provides fish and prawns, vegetables, coconut, coir, rice, and building materials. It feeds the ducks, whose eggs are often preferred to chicken eggs; and the water buffalo wallows in the coolness of the pond when the sun is hot.

Fish and Buffalo

Buffalo dung is used to fertilize the pond, small mounds of it being heaped at even distances all around the edge.

Buffalo dung is also used as fuel and as building material. It makes nice, dry floors. Women and children collect the dung from the neighborhood, carrying it in baskets on their heads.

Near Calcutta there is a mile-long wall, on the sunny side of a street, built for drying buffalo and cow dung. People bring their dung in the morning, shape round balls with their hands, and slap them on the wall. By the time the sun sets the round cakes are dry, and are then taken off the wall with a spatula.

Farmer Jamandre's Pigs

If a farmer has as many as four cows or buffalo, or eight pigs, there is yet another way of linking aquaculture and agriculture.

Farmer Ernesto Jamandre, near Iloilo in the Philippines, is the most successful fish farmer in the region. He has 120 hectares of fish ponds. He has laid out the ponds himself, with a canal at the head which distributes fertilized water to all. He stocks his ponds with four to five million fry every year, mostly milkfish. Farmer Jamandre also has 10 Brahman cows and 750 pigs. However, only eight pigs are needed for his special setup.

The pigs are housed in a sty just above the canal that feeds the fish ponds. It is a very clean sty. The pigs—very affable and personable pigs—are fed a variety of leftovers. Their own wastes are regularly whisked away and flushed down into a round tank that Farmer Jamandre designed and built himself.

The pigs' wastes are sedimented and fermented in the round tank. The sludge is used to fertilize the ponds. It nourishes blooms of the green alga *Chlorella*, which feeds the milkfish fry. Every three months the fry is transferred from the nursery ponds to the larger stocking pond in the rear,

Farmer Jamandre's milkfish farm, Iloilo, the Philippines.

Farmer Jamandre's pigpen. In the foreground: the round sewage fermentation tank.

Farmer Jamandre's pigs.

Methane gas burners in Farmer Jamandre's home.

where the milkfish reach market size and are harvested after another 65 days. Farmer Jamandre has a continuous, year-round harvest of well-fed milkfish.

Meanwhile, over the fermenting sedimented sludge, gas is forming in the round tank. Biogas. Methane. This the ingenious farmer conducts through a rubber hose to his simple home, where it lights two kitchen burners—there is enough gas to keep them burning twenty-four hours a day, with some to spare. Farmer Jamandre's next project is to install a refrigerator to utilize this excess energy.

Waste Recycling

Waste recycling is taking more and more sophisticated forms with the introduction of technology into aquaculture systems.

The Municipal Administration of Calcutta has successfully experimented with an integrated system that treats domestic sewage, feeds a fish farm, and fertilizes a vegetable plantation. Originally constructed by the British, the sewage plant was modernized by the Calcutta authorities in the 1950s. It served a farming community of 10,000 inhabitants which, in the meantime, has grown to 100,000. The capacity of the plant, however, is limited to a volume of 30,000 inhabitants.

The sewage is sedimented in a tank. The sedimented sludge is used to fertilize the vegetable fields: cucumbers and cabbages, eggplants, onions, peppers, and leafy vegetables, as far as the eye can see. The waste water is filtered through rocks and gravel and passed through a basin where it is agitated to remove detergents which form a foam that is skimmed off, while, at the same time, the water is aerated. Rich in phosphates and nitrates, the water is then channeled into the fishponds, where it produces luscious blooms of algae. The fish, carp, tilapia, and catfish receive no additional feed. In four months the fingerlings are fattened to market size.

Experimental facilities exist in Munich and Berlin, in the Negev Desert, and in Java. It has been observed by a number of scientists, however, that the integration of aquaculture into agricultural and industrial systems is little understood or practiced in much of the Western world, where high technologies often become frozen into administrative hierarchies. Aquaculture, agriculture, public health, and water pollution control are separate empires and it is still difficult, conceptually and administratively, to combine them in interdisciplinary efforts.

A sewage treatment plant, near Calcutta, India.

A sewage treatment plant and experimental fish farm, The New Territories, Hong Kong.

The fish ponds at the sewage treatment plant and experimental fish farm, The New Territories, Hong Kong.

For the economical use of waste water in aquaculture it would be essential to separate domestic sewage from industrial wastes, or to make industrial enterprises remove highly toxic material from their effluents.

The Water Hyacinth

In the nineteenth century, a British lady brought some water hyacinth from home and planted it in an ornamental pond which could be viewed from the terrace of her Raj mansion in southern India. The water hyacinth looked pretty, but it grew faster in the new environment than anything anybody had ever seen. Not only did it cover the pond with its lovely pale blue flowers and thick green foliage, it began to colonize ponds and rivers in the surrounding area. Grass carp and tawes, avid herbivores effectively in charge of weed control in many Southeast Asian ponds, did not take to the tough newcomer. The water hyacinth became a nuisance, choking Indian waterways.

More recently, however, it has been discovered by the National Aeronautic and Space Administration (NASA) in the United States that the water hyacinth, so far from being a nuisance, has the extraordinary capacity of absorbing, through its roots, mercury and cadmium, nickel and lead, even gold and silver from industrially polluted waters. What is more, these metals can be recovered from the harvested water plants. This can be done by placing them in a pit specially designed to prevent ground water infiltration. Over a period of years, the heavy metals in the pit accumulate to levels high enough to make extraction economical. NASA has also developed a sewage treatment process using water hyacinth. After purifying the sewage, the hyacinth can be fermented into biogas containing methane. One acre of sewage can feed enough water hyacinth to produce from 3,500 to 7,000 cubic feet of methane gas plus half a ton or more of fertilizer per day.

Aquaculture and Industry

The processing of fish, whether captured or cultured, inevitably produces waste material—trashfish, fish offal, body oil, shells—which can be reused.

Fish Products
Prawn culture in India now produces an annual waste of about 50,000 tons. Mountains of shells and heads issue daily from the swift hands of the processing girls on the prawn farm. By noon it is all swept away, some of it piled in pits where it naturally converts into fertilizer, some of it carted to the Central Institute of Fisheries Technology, where it is put to more sophisticated uses. The Institute has worked out a method of preparing a chemical called chitosan from these shells and heads which has wide and varied applications in the sizing of textiles, clarification of water and wine, and as a base for chromatography. High-energy food, fish soup powder, various pastes, fish meal, and fertilizers are extracted from the shell mountains. Nothing goes unused.

The freezing and canning of prawns, shellfish, and finfish has developed into a multimillion-dollar industry, partly labor-intensive, partly relying on highly sophisticated technology, for it must overcome a number of challenging problems related to the preservation of taste, texture, color, and control of bacterial flora.

Fish sauce is a product relished in oriental cooking, and in Bang Pla Soi, Chon Buri Province, Thailand, there is a fish sauce factory which mass-produces the sauce for worldwide export. Sardines and anchovies are unloaded from big trawlers. They are sorted for different types of sauce and immersed in brine for a month. It is rather awful-smelling slop. The slop is passed through various tanks, pipelines, and basins, strained and flavored, heated and cooled. The sauce is bottled and labeled. The wastes go into fertilizer production.

Fish sauce being bottled and labeled in the fish sauce factory; and barrels of fish sauce, ready for export. Chon Buri Province, Thailand.

In Moscow there is an enormous store called Okean where people can buy sauces, pastes, cakes, flours, oils, frozen and canned products. Outside the store films are shown about the nutritional value of the products, and how to prepare and serve them. Inside the store, Father Neptune himself stands in a corner with his trident, turning his head and flashing his eyes. He has a tape in his belly that bellows explanations and recommendations about the new products. People love it. They stand in line to get into the store, an hour before it opens. They listen, they sample the products, and they buy. Krill paste is really quite delicious, once you get used to it. Sometimes it is combined with caviar, which can now be produced synthetically from casein and other lacto-proteins, in a kind of cheese. Many people do their shopping early in the morning, on their way to work. There are refrigerated lockers where people can leave their packages, and pick them up on the way home in the evening.

The Industrialization of Farming

Not only waste recycling, processing, and marketing have been indus-trialized, however. The process of farming itself is being industrialized, for example, in Japan, in Hawaii, and in the United States. Fish farming and shellfish farming are now going through a stage poultry farming went through during the last few decades.

Eel farming in Japan, for instance, has become a highly energy- and capital-intensive industry. The annual production of eels in Japan during the past ten years has averaged 21,600 tons, of which 90 percent are cultured. More than half of the world production today is cultured. In Japan eels are often raised in hothouses where the water temperature, even during the winter months, is kept between 80 and 90 degrees. A waterwheel churns and aerates the tank water. Warm water, heated in an adjacent tank, is supplied to the rearing and feeding tank. A well-developed farm may have forty 20-by-50-meter indoor and outdoor rearing ponds. Approximately 40,000 tons of water are pumped in daily from a well or nearby river. Until recently, the young eels were fed on trashfish or, occasionally, frozen mackerel or sardines. They did a mag-nificent job of cleaning them up, leaving a bundle of intact skeletons. These were converted into fertilizer. During the last ten years, a synthetic diet for eels has been developed on a commercial scale. It is a mixture of fishmeal or silkworm pupae, rice bran, wheat flour, minerals, vitamins, and fish oil. There are special factories, highly mechanized and com-puterized, to manufacture formula feed for eels.

After about eight months of growing time, the eels are taken for processing to the eel farmers' cooperative. They are cut and skewered

Neptune offers advice to customers at Okean, the store in Moscow.

A pond aerator on an eel and Ayu farm.

A glass-eel trap, near Maisaka, Japan.

Opposite page:
Grading eels for processing on an eel farmer's
cooperative, Maisaka, Japan.

by machine and broiled en masse on a conveyor. Or they are sold fresh on the market and made into a host of delicacies, served with exquisite Japanese culinary elegance.

And yet even this highly mechanized culture is still linked to, and limited by, capture fishery. For the eel will not breed in captivity, and the mystery of his life cycle has eluded science. The eel industry must rely on the capture of "seed fish." At estuaries, between dusk and dawn during the cold winter months, the glass eels and elvers are captured with scoop nets, traps, tubes, and screens of all sorts. Since there are not enough of them, many are imported annually from Italy, France, the United States, and other countries. Induced breeding, however, is on its way. Eels have been successfully injected with pituitary gland extract, and it is only a matter of time before eels are spawned and hatched in tanks and the last barrier to mass production is removed.

4

EAST MEETS WEST

The science on which the miracle of the feeding of the multitudes is based comes mostly from the West: Woods Hole (Massachusetts) and Scripps (California), the Universities of Florida and Hawaii, to name only a few of the major "houses of science" in the United States; the Centre National pour l'Exploitation des Océans (CNEXO) in France; the oceanographic institutes of Hamburg, Kiel, Munich in Germany; the All-Union Research Institute of Marine Fisheries and Oceanography (VNIRO) in Moscow—to mention only a few in Europe. The research area is vast and interdisciplinary. It comprises all branches of marine biology and ichthyology, genetics, population dynamics, nutrition, pathology, environmental sciences, hydrology, agronomy and soil chemistry, management, sociology, and economics. These great institutions reflect the image of the House of Solomon or the College of the Six Days Works of Francis Bacon's *New Atlantis*, "to exhibit therein a model or description of nature, and the producing of great and marvellous works for the benefit of Man," and for the purpose of "the enlarging of the Bounds of Humaine Empire, to the effecting of all things possible." Like Bacon's House of Solomon, these institutions are "the eye" and "the lanthorn" of their realms.

These sciences, and the awareness of their importance, have deeply penetrated into the other parts of the world and become acclimated there. The Indian Government's Research Centre of Central Marine Fisheries with its various branches is a leading institution, by world standards, in research, training, and extension services. The Central Institute of Fisheries Technology in Cochin, India, and the Southeast Asia Fisheries Development Centre with its series of federated institutions in the Philippines, Japan, Malaysia, Singapore, and Thailand, are magnificently equipped and are pioneering in experiments with induced breeding, fish nutrition, and environmental studies. China employs millions of scientists, literally, in aquaculture research. Its prodigious production amounts to about 50 percent of the world's total aquaculture production.

In the East, however, unlike in the West, "aquaculture is culture," as one Indian scientist said. It comprises everything. It links everything. The philosophy underlying this systemic view of things is Eastern and it is very ancient. As a net is made of a series of ties, Gautama Buddha taught, so everything in this world is connected by a series of ties. "If anyone thinks that the mesh of a net is an independent, isolated thing, he is mistaken. It is called a net because it is made up of a series of connected meshes, and each mesh has its place and responsibilities in relation to other meshes."

A man named Ananda, Gautama Buddha taught, was given a supply of new garments. "What are you going to do with the garments?" the King asked. Ananda replied: "Oh King, many of the brothers are in rags; I am going to distribute the garments among the brothers."

"What will you do with the old garments?"

"We will make bedcovers out of them."

"What will you do with the old bedcovers?"

"We will make pillowcases."

"What will you do with the old pillowcases?"

"We will make floor covers out of them."

"What will you do with the old floor covers?"

"We will use them for foot towels."

"What will you do with the old foot towels?"

"We will use them for floor mops."

"What will you do with the old mops?"

"Your Highness, we will tear them into pieces, mix them with mud, and use the mud to plaster the house walls."

"Every article entrusted to us must be used with good care in some useful way, because it is not 'ours' but is only entrusted to us temporarily."

This philosophy, in various guises, is deeply penetrating Western culture in our day and is giving rise to what one might call the ecological world view. The ecological world view—like the ancient Eastern world views and unlike the philosophies that have prevailed for some millennia in the West—offers a new/old concept of man in his environment, man as part of his environment. It perceives the continuity between nature and man, between natural evolution and cultural evolution. It thus conceives technology not as antinatural but as part of nature. It views the earth as a whole—as dramatically photographed from outer space—whole and blue: more Planet Ocean than Planet Earth. It emphasizes continuity rather than discontinuity, integrity rather than fragmentation, the organic rather than the mechanical, cooperation rather than conflict as the main forces of an evolution which started billions of years ago in the oceans.

The Western concept of the common heritage of mankind comprising the resources of the seas and oceans, over which no state or person can exercise sovereignty or property rights but only temporary stewardship, has its counterpart in the teachings of Buddha.

That aquaculture has a philosophical base in the East and a scientific base in the West has far-reaching implications. In the East it is culture, it is life: culture to improve life by providing food and employment. It is embedded in the social and economic infrastructure. All that science can and must do is to make this culture more effective. In this respect, the East has much to learn from the West. In the West, aquaculture is science and technology, embodied in industry and providing profits: money. It has no social infrastructure. The economic infrastructure has yet to be created. In this the West has much to learn from the East.

It is on this meeting, this merger, that food for the world and peace for the world largely depend.

II
MEADOWS
AND FIELDS

◄ The Ulva harvest, Hong Kong Bay.

5

THE UBIQUITOUS ALGAE

There were forests and grasslands, deserts and rocks, before man took over and began to create new stone deserts, urban sprawl, spreading like a disease on the skin of the Earth; before he began to create new deserts in place of forest; before he began, ten thousand years ago to wrest fields and meadows from the wild by taming ten thousand weeds and trees and roots and bulbs.

The Earth is two-dimensional. Where there are deserts, there can be no fields and meadows; where cities sprawl, there can be no forests.

The oceans, occupying three times as much of the Earth's surface as the land, and as yet largely untrammeled by man, are three-dimensional, and their towering rocky mountains, their seemingly endless deserts may still be overlaid by wondrous wandering pastures, meadows, fields and forests, *Posidonia* prairies and *Caulerpa* lawns. The diatoms, in their myriad minute shapes, are the ubiquitous meadows of the oceans. The algae are the inventors of the process that converts carbon dioxide, water and nutrients into valuable edible substances which support the entire life system. Using energy from the sun, algae reenact the primeval drama of life's beginning by effecting the miraculous transformation of the inorganic to the organic.

There are thousands of different kinds of algae and seaweeds, ranging from the tiny diatoms to the giant kelp. Seaweeds are blue, blue-green, green, brown, and red. Some look like corals; like sponges; like primordial animals; as some primordial animals, in turn, are plantlike in appearance, the borderline between plant and animal, at this remove of space and time, not being so very sharp; some look like suns, like spiked spheres; like fans, like tubes; like pieces of skeleton; like birds' nests; like jewelers' artifacts; some look like miniature trees, their delicate branches moving gently in the wave; like cabbages, like lettuces, like shrubs; like thick forests of tall trees, where animals come for food and shelter.

Microscopic algae wander unattached, passively rolling on currents and waves or, like dinoflagellates, they propel themselves by their whiplike tails. The larger weeds attach by holdfasts to rocks or other suitable substratum offshore: anchoring devices, not roots. Roots draw nourishment from the soil. Roots are limiting. Algae feed through their leaves upon the constant nourishment of the surrounding waters. The trees of the ocean forests need no trunks, for to move with the currents and waves requires greater flexibility than to move with the wind, while the water supports the spineless growth more effectively than does the air. Not wasting their energy on rigid structures, the forests of the sea grow faster than anything on earth. The time required for growth of a green alga, *Chlorella,* is one day. At the day's end it redivides and becomes four; that is to say, one *Chlorella* will become four the second day, sixteen the third day, sixty-four the fourth day, until it achieves almost the same weight as the earth. Such growth, of course, has never happened, for the *Chlorella* could not get enough sunlight for the photosynthetic activity required to achieve it.

Even the Arctic, where ice floes are wearing down the shoreline and rivers carrying masses of melting snow cause great variations in salinity, is host to algal growth. Seaweed grows at a somewhat deeper level than in temperate zones. The waters of Iceland are one of the largest untapped seaweed sources of the world. The area covered with rockweed and kelp in the waters of Norway is as large as the area of cultivated land in that country.

On the other side of the Atlantic, Nova Scotia and the Gulf of St. Lawrence together form one of the richest seaweed areas in the world.

The seaweed beds of the Irish west coast are virtually inexhaustible. Kelp, Irish moss, and thongweed abound off the coasts of Brittany and Normandy. Autumn and winter storms throw masses of red algae, *Gelidium,* on the shores of Spain and Portugal. The graceful *Gracilaria* grows abundantly off the shores of the Atlantic bordering the southern United States. In the sunlit surface waters, plants may gain from 14 to 16 times their original weight during a 25-day growth period. The gulfweed, *Sargassum,* floats freely in massive quantities off Florida's Atlantic coast. The Caribbean, extremely rich in flora, has about 760 identified species.

The legendary Sargasso Sea, an area of some 5 million square kilometers bounded by ocean currents, contains a standing biomass of 11 million tons or more of the *Sargassum* weed, famous for its power to entangle ships. Little is known about the origin of this rapidly growing brown kelp. According to one theory it constitutes the remains of the flora of the sunken Atlantis.

Morocco is blessed with an abundance of red algae, including *Gelidium,* and around the Canary Islands, at halftide, the seaweeds form a flag pattern: a yellow fringe, formed by *Cystoseira abiesmarina,* is followed by a beautiful violet-red fringe of *Gelidium arbuscula,* followed by a wide dark belt, almost black, of *Gelidium cartilagineum.*

While the Mediterranean area as a whole is relatively rich in unused or underused seaweeds, the greatest accumulation of red algae in the world occurs in the Black Sea. The weed, *Phyllophora,* is found all along the shore of the Black Sea, but for some reason is most heavily concentrated in the northwest region, called Zernov's Phyllophora Sea. In the Southern Indian Ocean there are enormous quantities which could well yield a million tons or more of brown algae; around China, Korea, and Japan the wealth of weeds assumes fabulous proportions. Potentially the world's largest source of brown algae, the Eastern Central Pacific Ocean, which contains the giant kelp beds of California and northern Mexico, shelters a variety of other kelps and weeds.

The Southeast Pacific, particularly off the coast of Chile, is an area of intense productivity. Red algae are mainly found on the northern shores while brown algae dominate the southern waters. The Magellan Strait is one of the richest untapped seaweed resources in the world.

Even under the heavy ice cover of the Antarctic—2 to 3 meters thick—weeds grow from a depth of 100 meters. No one knows as yet how great this vegetative mass may be, but it is thought to be prodigious.

No one knows the total quantities of useful seaweed available for harvest around the world. According to one estimate (FAO) the world potential output of red algae is 2,610,000 tons wet weight, of brown algae 14,600,000 tons wet weight. At present about 400,000 tons dry weight (approximately 2,000,000 tons wet weight) are utilized.

Nori culture, Ise Peninsula, Japan. ▶

Preceding page:

The town of Sitangkai, Tawi-Tawi, Southern Philippines, the world's major supplier of cultivated *Eucheuma*.

Seaweed farmers setting out *Eucheuma* plants in Sabah, Malaysia.

Kelp, California.

Sunburst and kelp.

6

THE VERSATILE SEAWEEDS

Since time immemorial the inhabitants of coasts and islands have gathered seaweeds. In winter they raked them up from storm-battered beaches. They waded in the shallows, and gathered with their rakes what the storms had cut adrift. More daringly, they dived from boats and, with their knives, cut the fronds of the stipes. Wearing scuba gear has enabled them in recent years to stay below long enough to do a more comprehensive job on the patient weed, *Laminaria,* suitably called *religiosa,* or, if they should get caught in its snares, to disentangle without panic from the *diabolica*: for kelp can keep the unprovided diver captive till he drowns. Modern technology has made available a mechanical kelp-harvester, a barge whose bow is fitted with a specially built unit containing blades much like those used in agricultural mowers.

Many and various are the uses to which the seaweeds have been put throughout history.

On rocky islands, such as Aran Island off the west coast of Ireland, seaweed is mixed with sand to make soil. The mixture is simply laid on top of bare rocks: one layer sand, one layer seaweed, and so on. Potatoes, peas, parsnips, carrots, cauliflowers, and cabbages thrive where they could not otherwise survive.

Driftweed—seaweed or eelgrass that drifts to shore—is used as a fertilizer all over the world. Although relatively low in nitrogens and phosphorus, seaweeds are high in potassium and a large number of minor elements such as iron, boron, molybdenum, sulphur, calcium, magnesium. In Southeast Asia weeds are used to fertilize coconut palms. They are chopped up and buried in the ground in a ring around the tree. A brown weed, *Hypnea,* was used experimentally as compost for hibiscus in Cape Comorin in India. The plants showed a 73 percent increase in yield over plants that were treated with cow dung and ash. Fruiting reached its peak over a month earlier. In the Soviet Union, potato crops treated with seaweed fertilizer produced yield increases of 26 percent. In British experiments, wheat yields increased by 10 percent, grass by 16 percent, celery by 8, strawberry crops by 20 to 120 percent, and apple pickings nearly doubled. The world's biggest seaweed fertilizer factory has been built in Iceland. It produces several million tons of fertilizer a year.

Shepherds let their sheep graze on the weed-littered beaches of Ireland, Brittany, Norway, and Iceland. The sheep dogs, too, enjoy a seaweed snack; it makes their coats shiny. Many tests have been made of feeding meal of dried and ground seaweed to pigs, chickens and cattle. An admixture of about 15 percent to the normal diet for cattle, and of 10 percent to chicken feed has been found most beneficial. In a feeding experiment sets of twin cows were separated and herded in two groups. One group was given 200 grams daily of fortified rockweed (*Ascophyllum*) meal over a period of seven years. The total milk yield of the test group was found to be about 6 percent higher than that of the other group fed on identical rations except that the seaweed meal was replaced by 100 grams daily of a Norwegian standard mineral mixture. The cost of the mineral supplement was about the same as that of the seaweed meal. When all costs were calculated, the additional milk yield obtained by the seaweed-fed group of cows resulted in an increase in net income of approximately 13 percent.

Green algae such as *Enteromorpha, Cladophora,* and *Chaetomorpha* serve as food for milkfish in the Philippines. They are said to improve the taste of the fish. The algae are grown on twigs and branches of mangrove trees set in the water. A great number of blue-green algae and diatoms are also raised in various countries as fish fodder.

Throughout history and prehistory hungry humans have eaten seaweeds, and it is easy to predict an increase in consumption even in the now-overfed regions of the world. Seaweeds can be eaten fresh, tossed like garden salads. They can be stewed in soups, dried, soaked and boiled, or fried as a vegetable. The Burmese collect their weed, *Catanella impudica,* from rocks at low tide, dry it, dip it into hot water, and eat it with salt and prawn. The Indonesians cook *Ulva* in coconut milk. They also eat it pickled or smoked. The Russians have a tasty dish—available, in cans, at Okean—called "sea-cabbage," stuffed with mussels and rice.

Seaweeds can be made into bread: the Welsh make their famous laver bread with dried *Porphyra* (laver). The Indians of Baja California have for centuries gathered eelgrass, dried and ground it and baked it into bread. Some species of seaweed contain 300 times more iodine and 50 times more iron than whole wheat.

Dried seaweeds (*Porphyra* and a number of others) can be rolled into sheets: the national specialty of the Japanese, who use the sheets in a variety of ways—as covers for commercial crackers, as wraps for tasty morsels of raw fish, in soups or as a vegetable.

Unknowingly, we consume seaweeds in great quantities, in industrially prepared foods such as soups and sauces, mayonnaise, chocolate milk, sherbets and ices, cheeses and bakery products, fruit syrups and aspics, frozen foods and puddings, as thickeners, stabilizers, homogenizers and gels. The basic ingredients are agar, algin and carrageenin, the three most valuable phycocolloids (*Phycos* is Greek for seaweed, and *colloid* denotes the gluey, jelly-like substance of the product) extracted from *Laminaria, Gelidium, Gracilaria,* Irish moss, and a number of other weeds.

The list of industrial uses of seaweeds is far from exhausted, for agar is used in the preparation of microbiological cultures and in orchid growing, in photography, pharmacology and medicine. Dentists use it to make hydrocolloidal impressions—far more precise than wax impressions—and artists and criminologists use it to make casts of sculptures or of footprints, tire tracks, and so forth. Agar is used in storage-battery separation, fining of wines and juices, sizing of luxury fabrics, and as a lubricant for drawing tungsten, tantalum, and molybdenum wires. Algin, extracted mostly from the giant kelp *Macrocystis pyrifera,* is, amazingly, even more versatile. It can be used in foods in place of agar. In addition, it serves in the preparation of surprising numbers of pharmaceutical and cosmetic products—tablets (including aspirin compounds), suspensions, powders, toothpastes, laxatives and suppositories, hand lotions, hair conditioners and shampoos. The rubber industry needs algins for the production of natural and synthetic latex, automobile carpeting, babies' rubber pants, foam coatings, and tires. The textile makers use algin for the sizing of compounds for cotton and rayon, and print pastes and plastic laundry starches are made with algin. Algin can be spun into a fiber and woven into cloth. It can then be dissolved out in what is called the "disappearing fiber technique," which produces unusual patterns and textures. It can also be made into nonflammable fiber for children's clothes and camouflage netting for gun sites.

Algin makes wallboards stick, and paper bags, shipping containers, gummed tapes and decals; and who would suspect algin in pharmaceutical soaps and detergent packages, milk containers, butter cartons, insulation boards, food wrappers, grease-proof paper, and acoustical tiles? Or in paints, ceramic glazes, porcelain ware, leather finishes and auto polishes, boiler compounds and wax emulsions?

There are some other, minor phycocolloids that can be extracted from the algae: laminarin, a kind of kelp-starch; fucoidin, rich in calcium and sulfate; fuerin, a gluey material, long used in the Orient for sizing textiles; and there is an *alcohol,* mannitol, useful in the manufacture of explosives as well as in many of the items on the long list of phycocolloid products.

The list of pharmaceutical uses of seaweeds is at least as long as that of industrial uses. The antibiotic, growth-inhibiting, toxic, antiseptic, growth-enhancing, and other properties of about 14,000 species have lent themselves to a variety of pharmaceutical uses. *Ulva* makes a nice tea—known also as "bush tea" in the Caribbean—said to bring down fever and relieve high blood pressure. The Arabs have long used seaweeds to make laxatives. *Digenea simplex* serves as a vermifuge. Carrageenin has antipeptic or antiulcer properties and has been found useful as an anticoagulant and antithrombic substance. Certain species of *Laminaria* have been used in folk therapy in Japan for the prevention and treatment of hypertension and arteriosclerosis.

One of the most astonishing discoveries of recent years has been algin's capacity to absorb and eliminate from the gastrointestinal tract, radioactive strontium 90—the dreaded substance likely to be ingested by people drinking milk contaminated by nuclear fallout. Strontium 90, as is well known, causes leukemia and bone cancer in the victims of nuclear accident. Taken orally, algin reduces the uptake of strontium by a factor of 9. Even if the strontium has already reached the bone tissue, up to 25 percent can be removed.

The capacity to absorb noxious elements is not restricted to strontium. A red alga, *Lithothamnium calcareum,* is being used in France in the treatment of acid drinking water which otherwise corrodes the pipes. If filtered through the weed, heavy metal ions are eliminated through absorption or ion exchange.

John Ryther of Woods Hole Oceanographic Institution has been working for years on a project treating sewage effluent with seaweed. Nearly 100 percent of the nitrates and half of the phosphates from secondary liquid waste can be absorbed. Seaweeds are likely to play an increasingly important role in the development of "biological systems" which, in a more "organic" society, tend to supplement or supplant traditional mechanical or chemical systems to transform matter.

Thus the concept of an "energy farm" arises: something that is sea-farm and seamine at the same time.

Akira Mitsui, a Japanese researcher at the University of Miami, has identified a strain of blue-green algae that can separate hydrogen from oxygen in ocean water. Hydrogen is a waste product of their metabolism. So efficient is this alga that one millimeter of algal suspension will produce one milliliter of hydrogen gas in one day of 12 hours of light. If laboratory tests can be developed into a full-scale pilot project, all the energy of an average Florida home can be developed from an "algae farm" 8 meters square and 1 meter in depth.

Certain other algae concentrate uranium and can be genetically improved so that this concentration is even higher than in nature. Algologists are presently working on experimental "uranium farms," where uranium is concentrated by algae and extracted from them with a side product of methane and fertilizer.

However, the most ambitious project, bridging seafarming and seamining, is the Ocean Food and Energy Farm Project, carried out by Dr. Howard A. Wilcox and sponsored, in turn, by the U.S. Navy and various private companies.

After an initial phase, during which a 7-acre experimental farm is to be developed, the project is to be expanded over the next few years into a 100,000-acre farm, at the overall cost of 1.9 billion dollars. According to a detailed systems design, such a farm could yield enough food for about 750,000 persons and enough energy and other products to support more than 47,000 persons at today's United States per capita consumption level, or up to 300,000 at today's world average per capita consumption level. The farm products, according to the design,

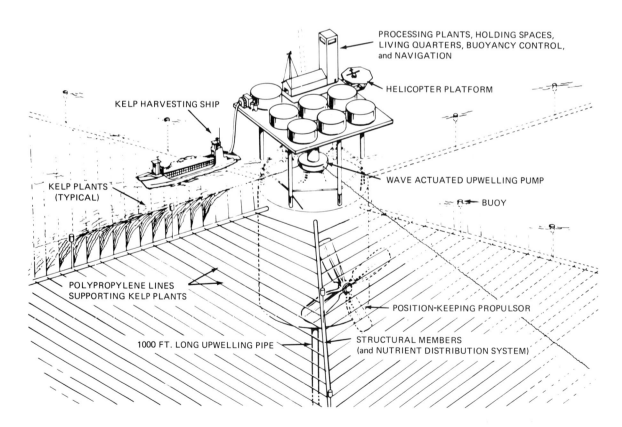

An artist's concept of a 1000-acre Ocean food and energy kelp farm.

would consist of feed for fish, livestock, and food for human beings, methane gas, fertilizers, ethanol, lubricants, waxes, plastics, fiber— indeed, a complete spectrum of useful petrochemical-type products.

The number of uses to which seaweeds can be put is increasingly important. "With the end of the fossil fuel era in sight," Dr. Michael Neushul said, opening the Ninth International Seaweeds Symposium at the University of California in Santa Barbara, "it is time to design a new world food and energy system geared to renewable energy. Solar energy input, prior to the fossil fuel era, was in balance with agricultural output. Today, it is not in balance, since fossil-fuel subsidies are used to increase agricultural yield to the extent that in some areas the population exceeds that which can be supported by solar-based agricultural yields. Thus it is certainly worthwhile to explore the potential of marine plants as collectors of energy and concentrators of nutrients, since these might contribute significantly to a solar-based food and energy production system."

7

SEAWEED FARMING

While the resource is enormous, and readily renewable, it is not always easily accessible. Often the richest kelp pastures are to be found on sea-swept rocks where no kelp-cutter can work. Remote, inhospitable areas in the Arctic and Antarctic may be rich in biomass, but it would be too costly and wasteful of energy to transport the harvest over thousands of miles to where it is needed. Where demand exceeds the supply of wild weeds, there is a need to increase the cultivation of weeds in seafarms. While the harvesting and use of marine plants is fairly widespread, however, their farming is by no means universal. Indeed, it is to the Orient that one has to look for mastery of the techniques of seaweed farming.

Basically, one can distinguish three stages of farming.

The first stage, which is also the oldest and simplest stage, is the prudent management of natural stocks.

The second stage, which can take a number of forms, is the manipulation of the environment.

The third stage, the most sophisticated and efficient, involves the control of the life cycle of the weed, or its full "domestication."

The prudent management of natural stocks may involve the regulation of harvest seasons and of harvesting techniques. Thus in Portugal, harvesting is limited to the second half of the year because the abundant agar-producing brown algae, *Gelidium* and *Pterocladia*, grow and reproduce during the first half. Seaweed harvesting in Portugal is traditional—everything in Portugal seems related to the ocean, from the placement of the castles to the Gothic ornaments, from the museums to the food—and the kelp harvesters dive from their small boats in traditional costume. The regulation of harvesting techniques is based on regrowth studies carried out in special seaweed reserves.

Canada, Great Britain, and Spain also practice prudent management systems for the exploitation of their natural stocks.

The second stage, the manipulation of the environment, can be carried out in at least four ways.

The first is to carry out ecological studies of the interaction between regional flora and fauna and the impact of human activities on both. An example of this stage of culture is the giant kelp culture (*Macrocystis*) off the coast of southern California.

In the 1940s and 1950s, the giant kelp was almost totally destroyed. The direct cause was determined by ecological study as overgrazing by sea urchins. This in turn was caused, on the one hand, by pollution, which encourages sea urchins, and, on the other, by the disappearance of the sea otter, overexploited for his valuable pelt. The sea otter feeds on urchins, which he pries open with a rudimentary stone tool while swimming on his back. Man had upset the ecological balance between otter, urchin, seaweed, and water, and an understanding of these interactions was needed if the kelp beds were to be restored. While it takes more time to replenish the stocks of the slowly reproducing sea otters, the urchins were eventually controlled by quicklime, which does not poison or pollute the water. Large areas were then replanted with mature kelp and young sporophytes (the asexual form of the plant). The reconstituted kelp forests provide food and shelter for many species of fish.

The second way is the artificial improvement and extension of the substratum on which the seaweed forest thrives and expands. In California this method is still in an experimental phase. In Japan, however, it has long been fully developed.

The Japanese began the culture of nori (*Porphyra*) as early as 1670 in Tokyo Bay. They threw tree twigs, split bamboo, or bamboo branches into the shallow, muddy water to collect the spores which settled on the branches and developed into thalli, that is, the edible part of the plant. This was done in September and October. After about two weeks the branches were moved closer to shore, preferably near a river estuary, exposing them to a flow of nutrient-rich, brackish water. They were picked or cut several times during the winter.

During the 1940s the Japanese introduced a further improvement: they replaced the twigs and bamboos with synthetic horizontal net beds stretched between bamboo poles. Production doubled. Some equally or even more efficient variants of this method were also developed in the 1940s: one consists of simple ropes stretched between bamboo poles—the so-called "hibi" method; another consists of nets fastened in upright position, perpendicular to the water surface; and yet another of floating nets, which, theoretically, frees the culture from limitation to shallow offshore waters.

Seaweed culture has two basic requirements: one is a suitable substratum from which to grow towards the sunlit surface of the sea, and such a substratum can be built anywhere and anchored in shallow water or kept in place by stabilizers at any depth. The second basic requirement is the presence of nutrients in the water. These are abundant in the

Rotating drums for the seeding of nori nets.

Rotary cutter for the harvesting of nori.

shallow waters of the continental shelf, but they are scarce in the upper layers of the deep oceans where they normally sink to the bottom and have to be brought up by natural or artificial upwellings.

The third way of manipulating the environment therefore is artificial fertilizing. To enrich the water of open-sea kelp farms by artificial upwellings, California is experimenting with pumps powered by energy from the waves, bringing cold, nutrient-rich water from a depth of 1,000 feet. This artificial upwelling system requires relatively little power, probably less than one horsepower per acre-foot of water per day.

The Chinese have perfected a different system of fertilizing their seaweed farms: they attach a number of cylindrical bamboo baskets to their raft cultures, each containing a porous clay cylinder with liquid fertilizer. The fertilizer gradually seeps through the cylinder and is absorbed by the young plants. Yet another method of fertilizing sea-weeds is being studied: an organic, rather than a chemical system, based on the large-scale introduction of nitrogen-fixing blue-green algae and bacteria into the Yellow Sea.

The fourth way consists in the modification of temperature and light at different levels of algal development.

The Japanese developed cold-storage techniques in the 1960s. After spore collection, nets are stored at deep-freeze temperatures, at which they can be kept for several months. They are then used as replacements for nets that have been harvested several times. The increase in produc-tion by this method has been substantial. Live storage of plants is practiced differently in different countries. The Chinese have developed elaborate techniques of controlling temperature, moisture, oxygen sup-ply, light, and microorganisms for the transport of the young weeds.

The third and final stage of "domestication" of seaweeds comprises the control of the reproductive process, artificial propagation of seeds or spores, and the genetic adaptation, selection, and improvement of the plant. This, obviously, can only be based on a thorough understanding of the life cycle of a plant.

"Mankind depends for food on 15 major crop plants," Dr. Neushul said at the Seaweeds Symposium, "these being the cereals, legumes, and the trees, sugar, and root crops. Almost all of these were first domesticated over 10,000 years ago by prehistoric agriculturists. In contrast, the domestication of marine crop plants is being undertaken by our own generation. There are now perhaps four domesticated marine plants, definable as such since the amount harvested from cultivated sources exceeds that taken from wild populations. These are the red algae *Porphyra* and *Eucheuma* and the brown algae *Laminaria* and *Undaria*. *Macrocystis*, the largest known marine plant, is still a wild plant, although sizable amounts of it are harvested and processed for the production of alginates."

Since the true domestication of marine plants has barely begun, its potential for the future of aquaculture may be very great indeed.

Although *Porphyra* had been cultured for centuries, no one knew what happens to the plant each year from the time it begins to rot and disappear in March until its reemergence seven months later. Then, in 1943, an English botanist, Elizabeth Drew, solved the riddle. So important was her work to the Japanese that they built her a monument near Tokyo.

What she discovered was that, as the *Porphyra* plants die, they release small spores. These are the result of sexual reproduction which takes place during the final stage of the plant's life cycle. The spores sink to the bottom and, in nature, settle on oyster or clam shells. There they germinate, forming threads which bore into the mantle of their host and appear as dark red incrustations, which had been considered by algologists as a totally different alga and was named *Conchocelis rosea*. In nature it grows during the summer months. When the days begin to shorten the alga forms a sack from which, in September, it releases its spores. Elizabeth Drew discovered that *Conchocelis* is not an independent alga but a phase—the asexual phase—of the life cycle of *Porphyra*. The new spores rise and attach themselves to rocks, twigs, or other natural or artificial substrata, where they grow into thalli which mature during the winter and renew the cycle.

The discovery of the life cycle of *Porphyra* revolutionized the industry, and made it possible to mechanize the collecting of spores. This is done outdoors, on the *Porphyra* farm, or indoors, right in the hatching tank. At the start of the nori culture season, farmers bring their own nets to the hatchery, and, for a small fee, are allowed to collect the spores on them. To maximize efficiency, nets may be mounted on one or two motor-driven wheels, keeping them in rotary motion to accelerate the settling of spores. The seeded nets are then taken to the farm and attached to supporting poles or rafts and allowed to grow.

The progressive domestication of *Porphyra* has led to an almost eightfold increase in production. Nori now is cultivated in almost all bays along the Pacific coast of Japan and in the Inland Sea—lovely to see, like vineyards transformed by a dream, bobbing in the blue, with the laborers in the vineyard moving about in little boats. About 50,000 families are engaged in nori culture.

The second weed that has been "domesticated" in the sense that it is artificially propagated and controlled during its whole life cycle, and that accounts for up to 15 percent of the total seaweed harvest in Japan, is *Undaria*, "Wakame." A brown alga, *Undaria* is a native of the colder waters of northern China, Russia, and Japan. Its life cycle is different from that of the red algae although, like *Porphyra*, it develops through alternate sexual and asexual phases. During its asexual phase, this alga has large fronds which are used for food. Usually, they are simply dried and chopped up and eaten as salad or vegetable. Recently, a salted product has come on the market and has become quite popular, greatly increasing the demand for Wakame.

In *Undaria* cultures, the spores are collected on "seedling twines" of synthetic fiber-spun yarn, which are wound around a square plastic frame submersed in a tank. There they are kept during the summer. In the autumn, the twines are taken from the tanks and suspended in the sea from rafts. By January they are ready for harvest.

A third seaweed that has been "domesticated" is *Laminaria*, in particular *Laminaria japonica*, a weed that naturally thrives in the subarctic waters of northern Japan but has been adapted successfully to the warmer waters down the Chinese coast. *Laminaria* is a favored dish in Chinese cuisine. It is also used for various medicinal and industrial purposes, and is therefore in great demand. The development of *Laminaria* farming during the last three decades or so is one of the great success stories of the Communist regime.

Until 1948, 40 or 50 thousand tons of dried seaweeds were imported annually into China. Production from natural beds in the northern provinces amounted to a meager 60 dry tons annually. Intense research on artificial propagation and raft culture of *Laminaria* began in 1952, and in the decade from 1949 to 1959 production multiplied about 400 times.

Through selection and genetic manipulation, a new strain of *Laminaria*, called "Hai-tsing No. 1," has been developed, which gives 20 percent higher yields and can withstand higher temperatures. Whereas in 1958 *Laminaria* culture was limited to the colder waters of the northern provinces of Liaoning and Shantung, it has been extended, thanks to this innovation, to Kiangsu, Chekiang, Fukien, and Kwantung Provinces. There is an extensive literature on the spore production, the selective processes and genetic improvement of the plant, the methods of shipping young plants, and the different techniques of raft culture and artificial fertilization.

The enormous Chinese expansion of *Laminaria* culture has been planned, systematic, well-integrated with industrial development. Demand, however, is still 10 times higher than supply, and an increase in production to one million dry tons per annum is considered realistically

Harvesting the kelp, Lienyun Port, China.

Overleaf:

The Tungshan Kelp Farm, Lienyun Port, China.

attainable through further intensification of production and geographic expansion. Given the demand and the extraordinary organizing capacity of the Chinese, a further tenfold increase is therefore not impossible. The importance of this lesson for the rest of the world cannot be overstated.

A fourth species that has been successfully domesticated is *Eucheuma*, one of the red algae. The cultivation of *Eucheuma* is in a way simpler than that of *Porphyra* and *Laminaria* in that *Eucheuma* propagates vegetatively, that is, without a sexual phase, simply by "cloning." In other words, a group of plants originate as parts of the same individual from buds or cuttings. Three different kinds of *Eucheuma* are utilized: *E. spinosum*, *E. cottonii*, and *E. striatum*. The most successful cultivation has been that of *E. cottonii*. Of this, one strain has been developed, called "Tambalang," which multiplies with extraordinary rapidity, one plant producing thousands of tons.

E. cottonii is cultivated particularly on small farms in the southern Philippines, in coral reef zones protected from storms, hidden paradises between the thousand islands. Two methods are used: net farming, which is somewhat more elaborate and requires a slightly greater investment in materials but is far more productive per hectare, and line farming. In net farming, a net is strung horizontally between 4 poles driven into the sea bottom. Small pieces of *Eucheuma* are attached with plastic strips ("tie-tie") to each of the rope intersections. They grow rapidly, and regenerate after harvesting, from 200 grams a piece to 1,200 or even 1,500 grams within 90 days, at which point they are harvested and regrow. Quite a bit of work is needed to keep the farm clear of drifting material and grazing sea animals. Line farming is the same, except that single polyethylene lines are used instead of nets, and production is less dense.

Eucheuma used to be harvested from the wild. The Filipinos eat it raw, as salad. In addition, however, it has considerable commercial value as a source for carrageenin. For this purpose it was overexploited for export to Japan and the United States. Without giving thought to conservation, gatherers grabbed whatever they could sell to the foreign companies. Nothing was left to grow, and exports declined dramatically. One company, Marine Colloids, Inc., of Rockland, Maine, in cooperation with the University of Hawaii, made great efforts to develop culture systems, train local growers, and even provide the materials to start farms. Thus, while harvests from wild stocks declined, harvests from cultures rose from 0 to approximately 10,000 wet tons and continue to increase by about 10 percent per year.

As so often happened, however, the operations by a foreign private

Two storage silos at the world's biggest seaweed fertilizer factory, Reykholar, Iceland.

company based on the extraction of a local resource by the people of a developing country were attended by a number of social and political problems and uncertainties. Benefits could be increased and more widely distributed if research, on the one hand, and manufacture of carrageenin and carrageenin products, on the other, were progressively nationalized by the Philippines.

Researchers of the Marine Colloids company stress that more research is needed to produce still faster growing strains with higher carrageenin yields. There is also an urgent need for studying the diseases of this alga, especially a local disease called "ice-ice," which wreaks havoc on the farms. Productivity per hectare can still be greatly increased.

Cultivation on the basis of vegetative propagation (cloning) has been attempted, and achieved in the laboratory, also for *Chondrum* (Irish moss) and some other weeds. The plants are detached and placed in an agitated tank or pond. This work is done under the auspices of the National Research Council of Canada and is expected to be commercialized and to yield large quantities of *Chondrum* over the next several years.

"Wet farming" in the past has been considered almost exclusively in relation to operations in the coastal zones or on land, as Dr. Wilcox points out. "If, however, a significant fraction of the earth's naturally received solar energy is to be converted to foods and other valuable products for man's use, wet farming will have to be extended to the vast areas of the open oceans." The technologies are on hand.

A great deal yet remains to be done to apply modern technologies to the expansion and intensification of inshore or nearshore cultures. Pests and diseases must be controlled; seed stock both for sexually and vegetatively reproducing plants must continue to be selected for the production of faster-growing, disease-resistant plants which are adaptable to a variety of environmental conditions and can be used for a number of purposes; artificial substrata, systems of fertilization can be improved; the costs of seed production, harvesting, and processing can be reduced. But there is no reason why cultures should not be expanded eventually into the oceans where there is less competition for space. Experts agree that the increasing world needs for fertilizers, animal feed, human food, energy, and alginates for industrial and pharmaceutical products will stimulate the culture of seaweeds worldwide. Seaweeds are highly efficient converters of solar energy. They are primary producers, noncompetitive with fish for nutrients, and therefore a suitable component of polycultures. Seaweed forests in the open sea would enrich the marine fauna. Seaweed cultures are immune to droughts or frost.

The oceans are multidimensional in their functions, and in each dimension as vast as any other.

To think of the oceans as meadows and fields is an awesome experience, like a journey to another planet, a Green Planet of Peace.

III
FARM
ANIMALS

◄ A mussel stake, Gulf of Siam, Thailand.

The fish
market,
Songkhla,
Thailand.

8

GETTING ACQUAINTED

Farm animals are creatures man has tamed and domesticated. He controls their reproduction, manipulates their genetic inheritance, raises them in total or semicaptivity, supplements their feeding or feeds them totally, has made their survival dependent on his husbandry, and modified their behavior accordingly. They maintain, in many cases, a childlike and dependent, often loving, behavior. We project human qualities on them, making them symbolize human virtues and failings—pride or faithfulness, brainlessness or cruelty.

The range of our land farm animals is wide, including insects, birds, and mammals which man has bred through the millennia for their honey or silk, for the eggs and the milk they yield, the meat and leather, the furs and feathers or wool, or the energy they furnish as beasts of burden or as herders or watchers or protectors, as hunters and retrievers, as man's best friends. Although they differ so greatly, from the handful of a chick to the majesty of an elephant, almost all of them are warm-blooded; most of them are vertebrates. With few exceptions they have lost the habit of eating one another. We understand their minds, or think we do.

Although the human mind has dreamed up creatures half-man half-fish since time immemorial, we do not identify with our seafarm animals in the same way. They are rather remote cousins. In fact, we feel so distant from them that we see them closer to the vegetal than to the animal realm. We "plant" and "harvest" our seafarm animals, and sometimes we measure the "crop" in "bushels." We have no personal relationship to them, although all this may change as seafarming is expanded and becomes more familiar: as aquaculture becomes culture and we know more about the mind of the deep.

Already there are fish psychologists, studying as well as steering the behavior of fish at sea, eliciting responses, communicating.

Thus young seabream have been trained to respond to certain sounds emitted by underwater loudspeakers. The sound is associated with the distribution of food. On the second day of an experiment and before any food became visible, several fishes appeared upon hearing the sound. After two weeks, all 800 fishes in the tank had learned to respond to the sound and hurried to gather around the speaker when called.

Schools of fish can be taught to respond to luring or threatening

One of the largest fish markets in the world, Manila.

sounds or, more correctly said, humans can learn to select luring or threatening fish sounds, reproduce them, and the fish will respond by approaching or fleeing. Thus the sounds emitted by yellowtails feeding in a cage were recorded. When reproduced from an underwater hydrophone, yellowtails from coastal waters would ascend to the hydrophone and search for food. Similar experiments were successfully carried out with mackerel, pink salmon, and squid. The control of fish behavior by the emission of threatening or luring underwater sounds appears to hold great promise for seafarm management. Application of techniques of underwater acoustics to aquaculture is a challenging new field.

Not only acoustics, however, is applicable. Visual aids also can be resorted to in conditioning and controlling seafarm animals—more effectively, perhaps, than land farm animals. As seafarming is coming into its own at a time when the behavioral sciences are far more advanced than they were in the early days of land farming and husbandry, there may in fact be some "phase skipping" in our relationship with our new associates.

Visual discrimination tests, applied in recent years to birds, dogs, horses and elephants, which made us marvel at their "intelligence" and their capacity to abstract and reason, are being applied to goldfish and squid and even marine worms. Amazingly, these animals with their rudimentary brains are able to learn to distinguish between different geometric shapes and sizes and to respond differently to different patterns. More puzzlingly, a trained marine worm will retain its learning even if its brain is subsequently surgically removed.

Other disquieting results have been obtained with the surgical removal of parts of fish brains.

Schools of fish have leaders, leaders who know how fast or how daringly to move without hurting or losing their followers. And we know where, in the fish brain, this sense of social responsibility is located. One fish psychologist picked an average fish from an average school and surgically removed that part of its brain where the sense of social responsibility is located. When the fish had recovered from the operation, it was returned to the school. It recognized no guidelines but darted around without rhyme or reason. But then a strange thing happened. After some moments of general consternation and confusion in the school, the operated fish was made the leader, and every fish followed in uniform motion, even to destruction, in the wake of a leader bereft of its sense of social responsibility.

Our seafarm animals are even more diverse in character than our land farm animals. They cover the wide range of the evolution of life in the oceans: bivalves such as mussels, clams, cockles, and oysters; crustaceans such as shrimps, crabs, and lobsters; and a great number of true finfish, of the saltwater, brackish-water, and freshwater varieties: dozens of kinds of carp and catfish, labyrinth fishes and pike and perch, black bass and striped bass and sea bass, sunfishes and snakeheads and murrels, sleeper gobies, whitefish and characins, frogs, mullets and milkfish and the true eels; salmon, trout, and sturgeon; shad, pompano, and yellowtail; snappers and groupers; plaice, sole, flounder, halibut, and turbot. We raise them for their eggs (caviar), their oil (cod-liver oil); we cultivate them for their shells (turtle, mother-of-pearl, pearls); we make them into fertilizers, soaps, cosmetics, or pharmaceuticals; we may also raise them as ornamental aquarium fishes, but mostly we raise them for their meat. Fish provide about 15 percent of all animal protein. This, however, is only part of the story. FAO statistics show that many more people depend on fish than on the other, more costly sources of animal protein such as meat, milk, and eggs. Only 20 percent of the world population consumes 70 percent of these luxury foodstuffs, while more than 50 percent of all people depend on fish for the greatest part of their animal protein.

By what criteria man selected land animal species for domestication we can only guess. Why did he tame the sheep and not the deer? Why the dog and the cat, and not the bear? How is one to explain the very special relationship between man and the elephant, which remains half-wild, half-domesticated?

The criteria for the selection of sea animals suitable for farming are, on the whole, easy to understand or to rationalize. A seafarm animal must be one that breeds, or can be induced to breed, in captivity. When the whole life cycle can be controlled, man may be able to improve upon those qualities in fish which are particularly useful to him, and the number of fish offspring will no longer be left to chance. In order for seafarming to be economic, a seafarm animal must grow rapidly and have a high "ratio of conversion" of what it eats into what it is. And it must be hardy, adaptable to various environments, and resistant to epidemic disease.

9

MOLLUSKS

Mussels

The mussel exists in practically all climates, and therefore comes close to being the ideal seafarm animal. It thrives in the cold waters of Brittany and the Netherlands, and in the tropical seas of southern India and the Philippines. The protein content of mussels is as high as that of beefsteaks, but steaks have 4 times as many calories and 18 times more fat than mussels. The farming of mussels, all experts agree, is the most efficient way to convert phytoplankton into nutritious and palatable human food. Mussels do this through a prodigious filtering activity. In the space of an hour one of these small creatures can filter up to 5 liters of water. The mussels on the ropes on a single raft filter up to 70 million liters per hour! Small wonder, then, that the mussels grow fast. The rate of growth, however, depends not only on the quantity of phytoplankton filtered but also on a number of other factors. The most important of these are light, current intensity, and water temperature. Mussels grow faster without light than with light. Mussels living in darkness became 26 percent heavier than those living in half-darkness and 70 percent heavier than those raised on the sunlit surface of the sea. As for the currents, the greater their intensity, the more water will flow past the mussel and the greater the speed of its filtering activity. Finally, water temperature has a very strong influence on growth. In the colder waters of northern Europe it takes from 3 to 4 years or even longer for the mussels to reach "marketable size," that is, a length of 6 to 9 centimeters. In Galicia, in northern Spain, it takes only 18 months for a mussel to grow to 9 centimeters, and in the balmy waters of the Indian Ocean the period from "planting" to "harvesting" is 4 to 6 months.

Although there are hermaphroditic mussels, most of them are either male or female. After reaching sexual maturity at 6 or 7 months, they "spawn," usually twice a year, by discharging their eggs and sperm into the water, where the eggs are fertilized and become free-floating larvae. One mussel may spawn 15 million eggs at a time, but most of these are never fertilized.

As the mussel larva grows its shell, it becomes increasingly heavy and seeks hard objects to which it can attach itself and become spat. Mussels

Mussel spat collectors, France.

attach themselves by means of strong, sticky hair called "byssus." Like barnacles they also attach themselves to the hulls of ships. In this way, it is said, the mussel, which is not indigenous to North America, emigrated from Europe at the time of Columbus.

The mussel's attachment to a hard object is not necessarily permanent. If mussels do not like their environment, they can detach themselves and crawl about, or propel themselves by a tiny jet stream, a gas bubble which they secrete; or they drift with the current until they find a more congenial environment, where they secrete a new byssus and attach themselves. This wanderlust can be a problem for mussel farmers.

Spat collecting is the most important phase of mussel farming, and a number of different methods have developed in different parts of the world.

Some of the best mussel farms in Europe are in Brittany, where the tides race at a speed of 60 miles an hour, carrying ample supplies of phytoplankton. What is good for growing and fattening, however, is not necessarily good for "planting," and the same tide that favors growth makes the attachment of the larvae or spat to the ropes very difficult. The Breton farmer therefore gathers his "seed" in southern France,

◀ Inserting an oak pole into the sea bed, France.   133

A fisherman in his boat working on a bouchet.

mostly near La Rochelle. There he hangs his ropes out over the natural mussel beds, and in a couple of weeks the spat attaches itself between the strands of the ropes. These ropes are then taken back to Brittany, wrapped spirally around wooden posts which are driven into the sea floor. The farmers do the wrapping either from a boat, when the tides are high, or at low tide, when they drive their oxcarts into the sea. They have to work fast, however, for the tide is faster than the oxcart.

This form of mussel farming was invented in the Middle Ages, by a sort of Robinson Crusoe, an Irish sailor shipwrecked near the coast of Esnodes, in Brittany. Like Crusoe he had no gun to shoot birds or anglehooks to catch fish to still his hunger. He was able, however, to fashion some kind of net which he stretched between wooden poles driven into the sea floor, and in this contraption he hoped to catch seabirds. The birds failed to comply, but a plentiful supply of mussels was soon found growing on the poles. The sailor also discovered that the mussels grew much faster on the poles than in their natural beds. Thus he became the first mussel farmer in Europe. His method of growing became known as "bouchet" culture. The origin of the word "bouchet," however, remains obscure.

In the Philippines, just as in France, the invention of the "bouchet" or

A mussel farming village in Thailand.

stake method was accidental: fishermen in the Bacoor Bay area discovered a great number of mussels attached to fish corrals. Now the stake culture of mussels, encouraged by the government, has grown to such an extent that there are more than fifteen hundred hectares covered by it in the Eastern Manila Bay area alone. Stakes, made of mangrove trees, bamboo, palm trees, or even concrete are driven into the sea floor. The bamboo has holes made in it so that it looks like a huge flute. The water enters the holes and prevents the bamboo from floating after it is staked.

The Thais also use the stake method profitably.

The village of fishermen and mussel farmers just behind the mangrove forest on the shore of Chon Buri Province is both picturesque and prosperous: its solid pile dwellings, which line both sides of the watercourse that serves as a main street, are well painted and even furnished with windowpanes.

The watercourse takes the mussel farmers in their motor barges through the mangrove forest and into the open sea, where a real "mussel wall" has been erected. A diver, clad in a white suit and headgear that leaves only his eyes and mouth visible, dives down to pull up a stake thick with luscious blue mussels. At sunset, the boat returns to the village, where the harvest is washed, marinated, and roasted over an open fire. During the months of algae bloom the mussels may be poisonous, but the poison is concentrated in the entrails; the rest of the flesh is perfectly edible. Extracting the poisonous part, which the Thais taught us, is like removing the stem and pit of a soft amarelle cherry. None of us got sick.

Near the farm there are mountains of mussels, which the women and children swiftly sort and shell. The lower-grade mussels are used as chicken feed, the shells go to the fertilizer factory. A mass of brightly colored, butterfly-shaped mussel flesh, spread on blue nylon screens, dries in the sun. This is a simple and effective way to store animal protein.

In other areas, where there is a sandy or rocky bottom, or where the sea is so deep that no poles can be driven into the sea floor, the spat-collecting ropes are hung from rafts made of bamboo, casuarina wood, or other timber and floated with empty oil drums, gas cans, or fiberglass-coated floaters. Spat collectors can be made of tiles attached to ropes, or of empty oyster shells, strung on nylon fishing lines through holes bored in the middle of the shell, or even old tires. After harvesting, the mussels are sorted and cleaned by shaking them in trays or baskets in the sea. The smaller ones can be returned to the raft and left a month or two longer to grow.

In an experiment in Calicut, India, 66 kilos of spat attached to ropes were hung from a raft in December. By April there were about 600 kilos of edible mussels.

The world's largest mussel producer is Spain, where the raft method has been perfected and industrialized. Spain produces almost one half—between 300,000 and 400,000 tons—of the world's annual harvest, and 95 percent of Spanish mussels come from the vast "floating parks" in the bays of Galicia, especially around Vigo. The ropes are lowered in April, when the spring spawning season reaches its peak. There is a second spawning season in the fall, but, strangely enough, spat does not attach to ropes in the fall. Farmers who need seed mussels in the fall must take them from natural beds and tie them to the ropes with a nylon net or stocking. The net disintegrates in the seawater as the growing mussels burst out of it.

Mussel farming in Spain is a family enterprise. Spanish experts say that production in the Galician bays could be increased fivefold. Such an increase might, however, have certain ecological ramifications. The indefatigable mussels on a single raft consume as much as 180 tons of organic matter in a year. Over half of this, or about 100 tons, is then excreted in the form of solid feces. The accumulation of this detritus on the sea floor might endanger the survival of other species. The problem

A mussel diver in the Gulf of Siam, Thailand.

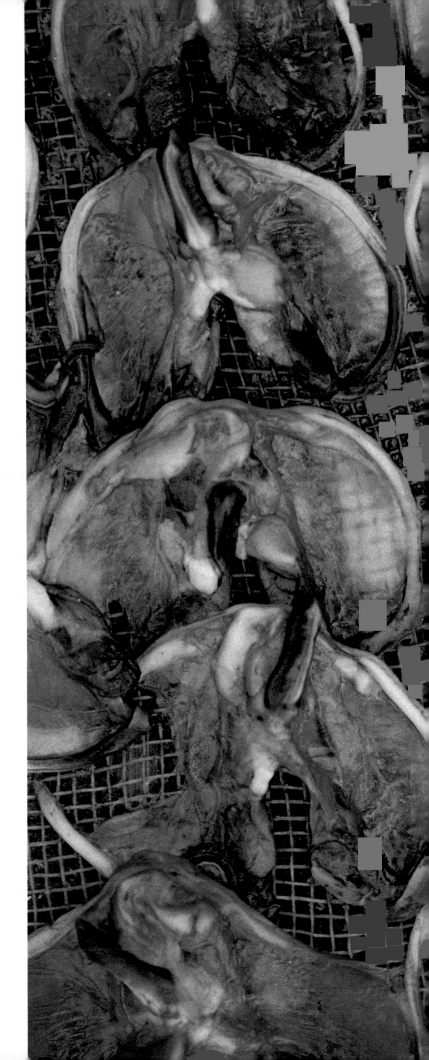

Mussels drying in the sun, Thailand.

Oyster raft culture, Ise Bay, Japan.

A large floating park, with the hull of an old barge as the center piece, Galicia.

has not yet been sufficiently studied. Introducing species in the area that thrive on the mussel wastes—for instance, sandworms—might be the answer. The sandworms could then be fed to fish, following the methods developed by John Ryther in his experimental polyculture at Woods Hole, Massachusetts.

The second largest mussel producer in the world is the Netherlands. Here, in the well-protected areas of the Zeeland stream and the Waddenzee, an entirely different form of mussel farming has developed over the past century and a half. Mussel seed is planted on the bottom on growing plots, thinned out whenever necessary, and transferred to deeper plots. During the winter months the mussels stop growing. The

Float equipped with a tall mast to hoist up ropes thick with mussels to be thinned out, or harvested, Ria de Vigo.

Industrias Maritimo Pesqueras, probably the largest plant in the world for cleaning, storing, and shipping mussels, Galicia.

small, family-size enterprises of 50 years ago have given way to large, industrialized enterprises. There are now about 12 of these in the Netherlands, producing about 10,000 tons a year.

Minor production takes place in England, Italy, Portugal, Norway, Scotland, Germany, and Yugoslavia. Also Latin America has made a beginning, especially in Venezuela and Chile.

And although North Americans normally do not eat mussels, there is now a surprising amount of mussel talk in the press: mussel poems, recipes, stories, and history. Fishery departments and Sea Grant programs are beginning to encourage the farming of mussels, and the *Wall Street Journal* mentions growing investments in the industry. It is safe to predict that Americans will go for mussels in a big way in the near future.

Oysters

Mussel culture is almost unique among aquaculture practices in the western world in that it produces a staple food rather than a luxury item.

The mussel's rich cousin is the oyster, whose culture, although in many ways similar to that of mussels, has been developed mostly in wealthy nations. The United States and Japan between them account for over 60 percent of world production, which currently amounts to 700,000 tons.

The sites on which to grow oysters have also become luxury items. The industrialization of coastlines and pollution from industrial and domestic sewage effluents have been destroying oyster beds at a rapid pace. In the United States one area after another has been closed to harvesting for health reasons. In New York State, which had been one of the main producers, oyster production has declined 99 percent over the past 50 years. But all this may be changing quite rapidly. According to the most recent forecasts, the demand for oysters will boost the present world production to more than 2 million tons by the year 2000. Such an increase would require the transformation of the traditional bottom cultures and of the more recent raft or other hanging cultures into mass production systems in totally controlled environments.

At present, however, traditional oyster farming is still remunerative in countries like France, where it has developed over the centuries into a fine and laborious art.

Two species are mostly cultured in France, the delicate flat oyster which provides the most expensive gourmet half-shell plates, and the Portuguese oyster which is somewhat more hardy but can, with some additional labor, be groomed as a luxury item as well.

143

A farmer stringing oyster-spat-collecting tiles.

A woman scrapes off young oysters for planting in the park, Morbihan, Brittany, France.

Harvesting oysters in a park in the Bay of Arcachon. The boxes into ▶ which the oysters are forked will be picked up by a boat as the tide comes in.

Oyster seed is collected from natural beds, on the southern coast of Brittany, mostly in the Gulf of Morbihan, and along the Southern Atlantic coast, between June and September. Oysters spawn when the water temperature is above 20°C. Sperm and eggs are released into the water by male and female individuals—if we may call them that, for they have no secondary sexual characteristics and produce sperm and eggs during different periods of their adult lives.

Fertilization takes place in the water. Free-swimming larvae hatch from the fertilized eggs, sometimes a few hours, sometimes a week after fertilization. After a few weeks they metamorphose and attach themselves as spat on a solid substratum.

During the winter months, the collectors are removed, and the young oysters, each one now about as big as a fingernail, are scraped off and "planted" in "parks," either right in the same area of southern Brittany, or they are sold to growers in northern Brittany. The parks, in harbors and estuaries, are protected by dikes of earth and brush. At low tide, the farmers, in rubber boots, with rakes and shovels, walk into the parks, spread the young oysters evenly on the bottom, and eliminate predators or fouling organisms such as barnacles, or algae, or excess sand.

When the young oysters are one and a half years old, they are gathered by hand and transferred to beds in deeper waters. There they grow for another two years. Then they are transferred back to shallow waters to be fattened and groomed for the market. This often takes place in so-called *claires,* that is, small, shallow, artificial ponds, especially and carefully prepared and fertilized. Some of the *claires* naturally produce a blue-green diatom which gives a greenish hue to the oysters, particularly relished by the French. The oysters double their weight during this "fattening" and "greening" period, which may take up an additional year. Thus French oysters are cultured over a period of four and a half years and moved three times before reaching the market. No wonder they are expensive!

While guaranteeing excellent quality, bottom cultures of this kind are becoming too wasteful of space and labor. Vertical culture, utilizing the three-dimensionality of ocean space, has a number of advantages over bottom cultures. First of all, the crop of oysters per unit area is greatly increased. Second, more food can be ingested by the oysters since they are more fully exposed to continuous currents. Consequently, the growth rate is about twice that obtained by bottom cultures. Third, the culture becomes independent of water depth and can be carried out farther away from the congested coastal areas, in clear and unpolluted water. And fourth, the oysters, kept off bottom, are out of reach of crawling predators.

Australian oyster culture.

Oysters are quite abundant along such tropical coasts as those of Cuba, Brazil, Sierra Leone, and Malaysia. In such environments they are often found attached to mangrove roots, but owing to overexposure to air at low tides and overabundance of fresh water during the rainy season, most of them die, or, if they survive, are stunted and unusable as food. By collecting their abundant seed, or by cutting the mangrove roots with the stunted oysters attached, and transferring them to a raft culture, an abundant protein source can be created for the inhabitants of tropical climates.

The heavy industrialization of oyster culture starts with the mechanization of seed production in hatcheries that have become veritable factories. This process has been perfected on a large scale in the United States and in Japan. In the United States oyster eggs were hatched in a laboratory at Johns Hopkins University as early as 1879. It was only in the 1920s, however, that the method was made commercially viable by W. F. Wells of the New York Conservation Commission. In the 1940s, his method, known as the Wells-Glancy method, was improved upon when two biologists of the U.S. Bureau of Commerce Fisheries Biological Laboratory at Milford, Connecticut, introduced induced spawning.

The selection of spawners, induced spawning, and hatchery rearing have paved the way for the selective breeding of oysters, producing races resistant to diseases, oysters that grow faster and develop better shell shapes, texture, and flavor.

Some 100,000 hatchery-reared Pacific oysters are now grown in containers in German coastal waters. The same area that used to yield one market-size oyster a year now yields 1,000 market-size oysters a year.

From the American West Coast hatcheries, packets of seed oysters arrive at Tap Pryor's spectacular Kahuku Farm right on the northern tip of Hawaii's Oahu Island. This system has been built around an abandoned aircraft landing strip which provides paving under and around the production trenches and algal reservoirs. At the seaward end of the farm, which stretches toward the mountains, seawater is pumped into a canal, by wind-power driven pumps. Still at the seaward end, the canal divides: about 80 percent of the water is conducted to the production trenches while about 20 percent is led into a tank, or "primary algal reservoir." Exposed to sea breezes, which are laden with phytoplankton seed, and heated by the tropical sun, the reservoir is a natural algae producer. The plankton blooms and multiplies, forming a rich green culture. This is then channeled from the smaller, primary algal reservoir to a much bigger secondary one, where the algal growth continues and expands. From there, the enriched water is returned to the seawater canal and flushed to the production trenches, long cement troughs about 3 feet deep, 3 feet wide, and 300 hundred yards long. The oysters are stacked on trays in the trenches, eight trays deep. As the oysters grow, they are thinned out, again placed in trays, and returned to the trenches. There are now 4 trenches, each producing half a million market-sized oysters a month. Production plans provide for 12 trenches, producing a total of 6 million oysters a month or 72 million oysters a year!

Since the whole production is land-based, no boats or divers are required. A flatbed tractor with a nuckle arm is used for harvesting. It can harvest 6 tons of oysters per hour.

The water from the production trenches, after feeding the oysters and flushing out their feces, is passed through a purification pond where a secondary crop of seaweed is raised. Then the water is returned to the sea.

Five species of oysters are used to produce pearls. Four of them are marine oysters, the fifth is a freshwater oyster.

Marine pearls are cultured on the southwest coast of Japan, in South Korea, and in China. *Pinctada fucata*, silvery white like the moon, is

Growing oysters on
an oyster farm near
Kahuku, on Oahu,
Hawaii.

Overleaf:

Abalone farming on
an oil rig, California.

Claires for fattening and greening of oysters, Brittany, France.

Production channels, Tap Pryor's oyster farm, Hawaii.

Oysters being sorted according to size, Tap Pryor's farm, Hawaii.

mostly used in Japan. *Pinctada maxima,* producing the largest pearls in the world, is cultured, above all, in Australia but also in Burma, Indonesia, and the Philippines. *Pinctada margaritifera* is the mother of steel-black shiny pearls and half-pearls and thrives in the Okinawa area as well as in Tahiti and Fiji. *Pinctada penguin,* called "Mabe" in Japan, generates large white pearls, while the freshwater oyster *Hyriopsis schlegeli,* cultivated in Lake Biwa near Kyoto and in some other places in Japan as well as in China, makes a salmon-pink pearl.

To produce a pearl takes time. It starts with the production of the mother oyster from spawn collected on cedar tree leaves and palm tree bark. This usually happens between July and September. Late in November, the young oysters are transferred to rearing cages hung from rafts or long lines. Ise Bay is crowded with such rafts, made of cedar logs or bamboo poles. Often during the winter, when the water is cold, the oysters are placed in heated water.

When the time comes, the oyster must be conditioned for the operation. Its gonad, which is where the pearl nucleus is to be inserted, must be free of eggs or sperm. In nature this period, immediately following the spawning season, is very short and the pearl farmer must extend it artificially by inhibiting maturation; this he can do by controlling temperatures and conditions of crowding.

The operators, clean and silent in their white coats, combine aspects

Trays holding oyster spat being hosed, Tap Pryor's oyster farm, Hawaii.

of the surgeon and of the watchmaker. The oysters are divided into two groups, the donors and the hosts. The donors have arrived at the end of their career, except that a piece of their mantle—the skin covering the soft tissues—will live on in the host for years to come.

The grafting of the mantle largely determines the success of the pearl production. The donor oyster is opened carefully with a sharp knife, and the animal is cut loose from its attachment to the shell. The main problem is to remove the mantle without injury. It is cleaned of all extraneous matter with the blunt edge of a scalpel and then cut into pieces 2 to 3 millimeters square. This graft tissue is kept moist with seawater, and at room temperature it continues to live for about two hours.

On another part of the long operating table, the pearl nuclei are prepared. Different materials can be used. Since it is advantageous that the nucleus should have the same or similar characteristics as the nacre deposited around it to form the pearl, the most suitable material at hand is the shell of the donor oyster. This is cut into small cubes which are then ground between two sheets of iron into rough spheres, with an average diameter of about 3 millimeters.

At one end of the long table, the most delicate part of the operation is performed. The host oysters are set in a shallow tray and covered with seawater. Within a few minutes the shells start to open, and bamboo wedges are gently inserted to open the shell wider and keep it open. Everything has to be done extremely gently, for if the animal is shocked or hurt, it dies. The opened oysters are inserted into an oyster clamp, and the operator smoothes back the folds of the mantle with a spatula, exposing the foot and body. Then an incision is made in the gonad, the graft mantle tissue is inserted, and the nucleus is imbedded in it and covered with a second piece of graft mantle. Thereupon the peg is removed and the oyster closes. Highly skilled oyster surgeons can insert from 2 to 5 nuclei in one oyster and perform this operation on 25 to 40 oysters in one hour.

During a period of convalescence, which lasts from four to six weeks, the oysters are kept in special cages suspended from rafts anchored in areas protected from winds and waves. At the end of the period, they are inspected and any dead shells are removed. If the operation has been skillful, the rate of survival is very high, about 90 percent. The cages are now transferred to ordinary culture rafts.

There they stay for another three to four years, during which the graft tissues begin to grow, forming "pearl sacks" completely covering the nucleus, whereupon the oyster begins to excrete the nacreous pearly substance around the nucleus that grows into a pearl.

Pearl oyster surgeons at work.

A master operator inserting the pearl nucleus.

An oyster in an oyster clamp, Ise, Japan.

At least three times a year the cages are lifted and cages and oysters are carefully cleaned. They are also frequently moved—sometimes over hundreds of miles—to ensure the best environmental conditions first for the growth of the oyster, then for the growth of the pearl. When the pearl is ready, the oyster will have had a long and well-protected life, with rather less suffering than is inflicted on its kind by nature.

Until quite recently the Japanese maintained a virtual monopoly of the technology of culturing pearls. Australia started its pearl culture in collaboration with Japan in 1956, and now the Philippines and Burma have similar arrangements with Japan. Cultured pearls are also produced on a small scale in Hong Kong, Palau Island, Celebes, and a few other islands in the South Pacific. Pearl cultures outside Japan are usually joint ventures in which the Japanese partner supplies the technical staff and is responsible for marketing the product, while the host country plays a relatively minor role in establishing and maintaining the farm.

India is probably the first country to establish an independent pearl industry. Fishing for natural pearls in the Gulf of Mannar and the Gulf of Kutch has a very long history. The pearls of the Gulf of Mannar were one of the earliest items in India's trade with Rome, Egypt, Greece, and China. But the natural beds have been exhausted. During this century, production has been negligible, and for the last twelve years there has been no pearl fishing. The first cultured pearl of India was produced in 1973, and since then the industry has developed quickly. Oysters and pearls grow much more rapidly in the waters of India than they do in the colder waters of Japan. Besides pearls, the Indian pearl industry produces food: the pearl oyster meat is edible. In the system of Indian medicine—as in Chinese medicine—pearls are used in the cure of many diseases, and the shells are used as fertilizer, ground up as poultry feed, or fashioned into ornamental articles.

The recent sharp decline in pearl production in both Japan and Australia, due to water pollution and labor problems, puts India, with its warm and relatively unpolluted water and abundant and inexpensive labor, in a good position to compete in the established pearl culture market.

Clams, Cockles, Abalone

Clams and cockles have also been cultured through the ages in Spain, Portugal, Japan, China, and most of Southeast Asia, providing food that is inexpensive, tasty, and rich in protein. The United States has now taken the lead in modernizing traditional methods of farming clams, especially the hard clam (quahog) which is cultured in New England and Long Island on a large commercial scale. In the cold waters of this region, quahogs take 5 to 8 years to grow. Therefore, there is now a tendency to move the industry southward. The world's first commercial clam hatchery, with a projected production of 300 million seed clams per year, has been built in Virginia.

The culture of shellfish is being combined with other uses of offshore water. Thus, abalone cultures are being developed on offshore oil rigs which, it is claimed, provide an ideal environment for their growth. These projects are still in an experimental stage, however, and looking at the oily waters round the rigs, the outcome of the experiments does not seem too secure. The abalone are hatchery-reared and transferred to cages hung from the platforms (rafts would not be significantly more costly) and fed chopped seaweeds and plankton. The coupling of aquaculture with the utilization of hot water from electric power plants seems economically rewarding. This is being done, on an increasing scale, in northern Japan, where low water temperatures in winter slow the growth of cultured abalone. The shell length of cultured abalone larvae increased more than three times as fast as naturally grown larvae. The high water temperature increases food consumption and conversion in young abalone. (It also increases the propagation rate of the diatoms upon which the abalone larvae feed.) Shrimp, yellowtail, and sea bream have been cultured in heated discharge water, with similar results. This is done by simply placing floating sea pens in the discharge water outfall. It is economical and can be applied on a large scale.

10

CRUSTACEANS

Shrimps

A few years ago, a woman was haled before a court in London. She was charged with cruelty to animals. She habitually amused herself by roasting live shrimps on a low-heat burner, making them twitch and jump and protracting their agony. The court refused to convict her and sent her home with a mere admonishment. The members of the jury were unable to agree whether shrimps were "animals" and what constituted "cruelty" against them.

The jurors never saw a live shrimp! They never watched it suddenly stand up, raise its head as if surprised by something, roll its strange goggle eyes and grind its two oddly human large molars. They never saw a shrimp arch and stretch, cast off its shell, and, lie down next to it, exhausted from the labor of molting.

The female shrimp molts for two purposes: in order to grow and in order to mate. The female shrimp undresses for her wedding night.

Shrimps have an exceedingly complex anatomy and a life cycle to match. After the egg hatches, the shrimp passes through a great number of metamorphoses. During each phase it has a different aspect, different habits and different nutritional needs. This complicates scientific farming. Also, shrimps are a rather quarrelsome lot, frequently injuring and devouring one another. Shrimps are cannibals.

The shrimp family is a very large family. Its members include the tiny krill which crowd the icy wastes of the Southern Ocean and the huge, proud-looking tiger prawn of the tropical Indian Ocean. Shrimps live in practically all water temperatures and in all degrees of salinity. The large, long-armed Malaysian prawn *Macrobrachium rosenbergii* spends most of its life in fresh water. It is intensively farmed in many parts of the world, including the United States where culture systems for mass production have been designed in Hawaii and South Carolina.

During the past three decades, however, marine shrimp farming has been successful mainly in the countries of Southeast Asia and Japan. It is usually a very simple process, and it yields good returns in nutrition and cash.

In Thailand, for instance, most shrimp farms are within 5 kilometers of

Thais make a delicious shrimp cracker out of a shrimp paste which is rolled into sheets and dried in the sun.

The research facility for genetic studies on *Macrobrachium,* Dr. Fujimura's Institute, Hawaii.

the sea and are connected with it by canals. The ponds are bounded by earthen dikes and open into the canal by means of one or two sluice gates. More than half of the farms use a pump or traditional dragonwheel to introduce the seawater into their ponds. Often one sees even simpler devices in use: two men or two women swing a wooden container which scoops water from the canal and pours it into the pond. They swing it for a long time.

On other farms, the tides enter naturally into the pond, carrying young shrimps with them. The shrimps settle at the bottom of the pond and grow rapidly. After four to six months they reach market size and are harvested in a net that is dragged over the pond floor. There is no artificial feeding and the only work done is the maintenance of the dikes and sluices and the elimination of predators that occasionally stray into the pond. The yield—of white and pink shrimps—is relatively low: a mere 109 kilograms per hectare on the average for Thailand. This, nevertheless, constitutes about 70 percent of the income of the Thai shrimp farmer, the remaining 30 percent being made up by sea salt and paddy farming or marine fishing and trading.

Tidal ponds at an experimental farm in Kakdwip, in the Ganges Delta, India. Tiger prawns, five kinds of carp and mullet are grown in polyculture.

161

These methods of farming are common to all the countries of Southeast Asia, from India through Indonesia, from Malaysia to the Philippines.

The Filipinos have improved their system of farming giant tiger prawns. Instead of waiting for them to drift into the ponds, they go out and catch the fry with nets from boats or with traps made of bundles of branches and twigs, near the riverbanks. Their growing ponds are quite complex in construction, subdivided into a small nursery pond and a large rearing or production pond. Sluice gates permit the passage of tidal water, filtered through a nylon net screen to keep out predators, and the ponds are equipped with drainage ditches to facilitate harvesting and cleaning. Before stocking, the pond is drained, allowed to dry, and treated with agricultural lime. It is then partly filled with just a little seawater. The shallow water, heated by the sun, develops an intense algal and planktonic growth called lab-lab, which serves as food for the shrimp fry. Chicken manure is often piled on the edges of the pond as an additional fertilizer.

The shrimp fry are stocked at a very high density: up to half a million fry per hectare, in the diked-off nursery pond. At some farms the fry are given additional food, usually rice bran and fish meal.

After a few weeks the nursery dike is opened and the fry are dispersed in the larger production area, at a density of about 10,000 per hectare. The supplemental feeding is usually continued in the production pond. The shrimp feed by night and burrow into the sandy bottom by day. They are kept in the production ponds from 6 to 12 months, until they have reached market size. Harvesting is done by draining the pond and catching the shrimps in bag nets or bamboo traps.

For the farmer who depends on naturally produced shrimp fry, however, there are good seasons and bad. Sometimes the shrimp larvae riding the tide through the wooden sluice gate or trapped by the riverbank are plentiful; sometimes the bulk of the hatchlings, for one reason or another, have perished at sea. The continually increasing demand is a stimulus to research in artificial production.

Shrimps will not breed in captivity.

In some farming systems, pregnant females are caught in the wild. In other systems, maturation of captive females is brought about artificially.

The male shrimp is not like the female in his reaction to captivity, and no special treatment is needed to induce him to mate and deposit his sperm in the seminal receptacle or spermatophore of the female.

No matter where she came from and how she matured, the impregnated female is placed in a spawning tank. She spawns by night, swimming about vigorously and emitting clouds of spawn: sometimes as many as 300,000 to 400,000 or even a million eggs at a time. The "spent" spawner must be removed at once, lest she devour her own offspring.

The newly hatched nauplii are fed after two days, when they have consumed their supply of egg yolk. The feeding of the tiny baby shrimps is a rather elaborate affair, the result of lengthy research, and much further research will be needed.

After six moltings, the nauplius is transformed into a zoea. At this point it is fed on a culture of diatoms and zooplankton. *Chlorella* is cultured for this purpose, and the tiny brine shrimp Artemia. After three more moltings the zoea, a roundish transparent being with wiggly legs, is transformed into a mysis, with a tail that gives the animal the beginning of a shrimplike appearance. Bread yeast is added to the diet and more zooplankton, especially rotifers. After three more moltings, the mysis becomes a postlarva. It now looks like a complete little shrimp and thrives on a meal enriched with minced tuna meat and ground trashfish. Research in Japan and other places indicates that the growth rate of cultured shrimp or other crustacean or finfish is highest when the amino-acid composition of their feed is just the same as that of their own protein: perhaps the most convincing explanation for cannibalism in general. At any rate, the amino-acid composition of kuruma-ebi protein—kuruma-ebi being the large shrimp preferred by the Japanese—is closest to that of shortnecked clams, squids, and some smaller shrimps. These, therefore, make the most suitable artificial feed. Steamed soybeans and wheat gluten are added and the whole mixture is put through a mincer, placed in a trough, and steamed. It can be frozen for storage. The young shrimp eat about 10 percent their body weight daily. The water in which they live must be well aerated, and about 20 percent of it must be replaced every day.

When they are about four weeks old, the young shrimp, or postlarvae, are ready to leave the nursery and to be shipped to wherever their rearing process is to be completed. For short trips they can be placed in a small tank, at a density of half a million individuals in one cubic meter. Soon, however, they start devouring one another. This tendency can be checked to some extent by lowering the water temperature. For longer trips they are packed in oxygenated polyethylene bags, about 5,000 to 6,000 in 6 to 8 liters of fresh seawater and as much oxygen, and the bags packed in boxes. The boxes are loaded on refrigerated trucks or trains or planes with a temperature maintained at 18°C. The shrimp can travel comfortably for 12 hours this way. Another popular method is to place them in a live-fish hold on the bottom of a fishing boat, with a screened opening so that seawater flows through as the boat moves.

Wild shrimps are still abundant in the waters of the Caribbean. There, however, the harvesting by trawler is increasingly difficult. Some fisheries of Gulf shrimps have been practically eliminated by the drying up of tidal swamps and estuaries. Furthermore, rising fuel costs account for diminishing profits. To bring in the maximum possible amount of the high-priced commodity, commercial shrimp trawlers have generally adopted the practice of simply throwing overboard the inadvertent catches of less valuable fish. On the average, 10 tons of fish are thrown overboard to produce 1 ton of shrimp. In a region where animal protein is in short supply, this practice is looked upon with increasing indignation by native coastal populations. Lastly, the new Law of the Sea is bound to upset existing fishing practices, and until fishery zones are determined and fisheries agreements are renegotiated, the trawling business is likely to have a recession. All this encourages farming.

In Argentina, "langustines" and other shrimp species indigenous to the waters of the region are raised in tanks and indoor and outdoor pools and fed a diet of squids and crustaceans supplemented by mineral salts and vitamins. Mexican experts have developed shrimp cultures, partly financed by United States capital, which has also backed some recent developments in Ecuador and Honduras.

In Africa, shrimp farming figures in the National Plan for Development of Aquaculture in Egypt (1975) but has remained at an experimental stage; in Nigeria, its development was interrupted by the civil war.

In Europe, a considerable amount of research is carried out by the French, but Germany and the United Kingdom are closer to production on a commercial scale.

In the United States the prospects appear to be good. A Japanese-American owned farm in Florida, Marifarms, Inc., has leased 1,000 hectares of estuary from the state and assumed, as part of the agreement, the obligation of releasing annually 20 million artificially raised shrimps back into the sea. Experimental work with white, brown, and pink shrimps, in natural embayments, ponds, and tanks has been carried out in various places in Texas and Florida.

Despite the absence of large-scale development in large parts of the world, shrimp production is by far the most important aspect of the farming of crustaceans. But other crustaceans also have been cultured more or less successfully.

Crabs

Experiments with blue crabs and king crabs have been undertaken at the University of Miami as well as in Japan. The capture of egg-bearing females, their placement in hatching chambers, the nursing of crab larvae in net cages and their feeding with brine shrimp, shrimp juice, clam, and seaweed (*Laminaria*), does not seem to pose any particular problem. The commercial incentive is strong, for the king crab fishery, one of the high-value fisheries of the North Pacific, has almost entirely collapsed. The difficulties, however, lie in the troublesome character of the little beasts. Their cannibalistic tendencies are worse than the shrimps'; once they reach postlarval stage they begin to attack one another relentlessly even under optimal spatial conditions. The farmer may attempt to distract them with lights or abundant food supplies; nevertheless, they are likely to destroy and devour one another.

Lobsters

Attempts to farm lobsters—both the big-clawed North Atlantic varieties (*Homarus americanus* and *Homarus vulgaris*) and the spiny lobsters of the Pacific or Mediterranean—date back to the early years of this century. From a biotechnical point of view, again, there are no insurmountable difficulties. But the lobster is the elephant of the crustaceans. The time of its growth and maturation is long indeed.

Like all female crustaceans, the female lobster molts before mating. When molting, she apparently emits a pheromone or scent which attracts the male. A-courting he comes on the tips of his walking legs, swaying from side to side. When he reaches the female, there is a scene of great tenderness. The two lobsters caress each other with their antennae for as long as 20 minutes or half an hour. It could be a scene from a science fiction film depicting love among extraterrestrial beings. After the antennae scene, the male lobster gently rolls the female over on her back, and mating takes place. The female stores the sperm for about a year, after which time she spawns her eggs, fertilized, and glues them to her swimmerets—a pair of legs specially adapted for swimming or carrying eggs. There she carries them for another year or so, before hatching. After hatching, it takes the young lobster six years to reach sexual maturity and commercial size (half a kilogram). The life span of the lobster is between 50 and 100 years, during the course of which it may reach a weight of 19 kilograms or more.

This natural biorhythm may be accelerated by human intervention and manipulation of the environment. The time between mating and hatching can be halved by controlling the water temperature, and experimental heating of the water has reduced the normal growth period to two years. The same result was achieved by raising lobster larvae in the tropical seas around India and Hawaii.

165

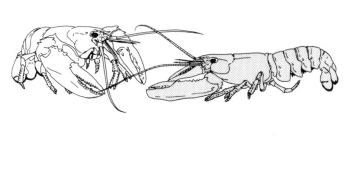
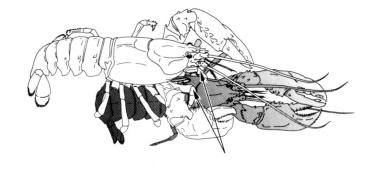

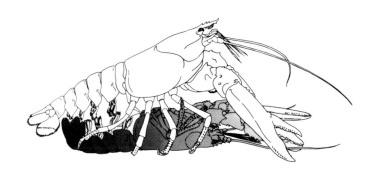
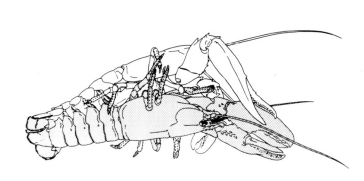
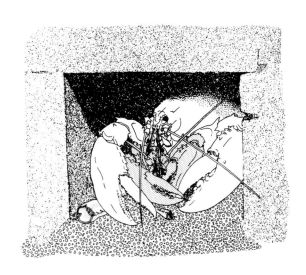

Lobsters mating.

Until recently, not much was known about the dietary requirements of lobsters. Studies undertaken in the 1970s indicate the usefulness of adding codliver oil to the diet.

Lobsters grow just after molting. Thus the more often they molt, the faster they grow, and ways and means to shorten the time between molts are being studied in various places. Canadian scholars have experimented with bilateral eyestalk ablation. Attempts in the 1960s were not encouraging, however; many of the blinded lobsters died. But in more recent experiments in Halifax, Nova Scotia, ablated lobsters gained over ten times as much weight in the year following the operation as normal ones.

Louisiana crayfish culture, New Orleans, Louisiana.

Another difficulty to be faced by the lobster farmer is a lobster disease called gaffkemia. Gaffkemia is a bacterial infection, transmitted through breaks and ruptures in the lobster shell. Since lobsters are as violent as they are tender and frequently inflict wounds on one another, the disease tends to spread. Infected animals become lethargic and weak, and this, in turn, stimulates the cannibalistic tendencies of those around. Before dying, the victim releases masses of bacteria which penetrate the many wounds of the cantankerous cannibals. If an epidemic occurs, whole cultures may be wiped out. But there are ways to cope with the disease, both preventive and curative, and forecasts for lobster culture on a commercial scale are not too pessimistic. The increasing scarcity of lobsters and their high market value may help make the research efforts yield commercial results in the near future.

Crayfish

A large family of small, lobsterlike creatures, the so-called crayfish or crawfish, live in freshwater swamps and shallow ponds in most parts of the world. In Louisiana, the culture of crayfish has been imported as part of that state's French cultural heritage. Close to 7,000 farms are devoted to this form of aquaculture. In addition, rice fields are utilized for crayfish culture in two-year cycles of crop rotation. Production yielding a return of $10 million annually to crayfish farmers in the state has been achieved.

Crayfish make eminently suitable farm animals. They are hardy and environmentally adaptable. At least, the American varieties of red and white crayfish are; crayfisheries in other parts of the world, for example, in Finland, have been wiped out by epidemics and plagues. Crayfish are modest in their food requirement: though omnivorous and partial to morsels of fish or beef, they thrive on the natural growth of rice stubble, alligator grass, water primrose or duckweed, and generally no artificial food is added to this diet. Also, they mature and reach market size in one short season. The farmer "plants" young adults in late spring or early summer, this being the mating season. Soon after mating, the females start burrowing holes in which the animals, usually in pairs, estivate. When the burrowing starts, the farmer drains his pond, weeds it, if necessary, and attempts to eliminate any predatory fishes. The crayfish eggs hatch in September, and at that time the ponds are reflooded. In November or December the first batches are ready for harvesting and the harvest continues until the pond is drained the following summer. Harvesting is done by luring the animals into meat-baited traps distributed throughout the pond.

A harvest of shrimp
on a prawn farm,
Kahuku, on Oahu,
Hawaii.

The Indian spiny lobster, Vizhingam, Madras, India.

A tiger prawn, the Kakdwip experimental farm, West Bengal, India.

A shrimp fisherman casting his net, Narakal, India.

Baby eels feeding, on an eel ▶
farm, Maisaka, Japan.

Netting sturgeons on
the Volga.

Drying catfish, Uthai
Thani Province,
Thailand.

Smoking catfish, Uthai Thani Province, Thailand.

Unloading a catch of shrimp, Songkhla fish market, Thailand.

Drying cuttle fish in a village in Southern India.

A milkfish harvest, Laguna de Bay, the Philippines.

11

FARMING FINFISH

Swift herds of true finfish are husbanded, from the semiwild to fully domesticated stages, in tubs, tanks, raceways, ponds, lakes, semi-enclosed seas and wide oceans. The field is vast and is rapidly reaching parts of the planet where fish farming is new and untried.

Some American Experiences

Thus, in the United States, for instance—not a fish-farming country by and large—the area of catfish ponds increased from practically nothing in 1960 to about 50,000 acres by 1976. The bewhiskered catfish, farmed extensively in Southeast Asia and other parts of the world, are peculiar animals. They are air-breathers, which means they can be kept at great density in oxygen-poor water. Packed in boxes without water, they can travel considerable distances, provided their skin is wet. If, on the other hand, they cannot reach the surface of their pond to breathe, these fish may "drown." It is the male catfish who minds the eggs, seeing to it that they get enough air by fanning them with his fins.

The economic value of the new industry is quite considerable. In the county of Lonoke, Arkansas, in 1965, one-twelfth of an area that had been planted with soybeans was turned over to catfish culture, and the next year yielded almost as much as the total soybean area!

Canadian estimates indicate that prairie ponds, stocked with trout, are 2.5 times more lucrative per acre than wheat farming.

Trout farming can become a bio-industry of incredible intensity. In this field the United States has achieved a record, with the world's greatest trout factory, the Snake River Trout Company near Buhl, Idaho. Here one farm alone, Bob Erskin's establishment, produces 600,000 kilograms of trout every year—or 12 percent of the total United States production—on 10 acres of land! Apart from a few buildings, grouped, mostly, at the end of the farm and including a hatchery and a processing plant, the farm consists of an impressive array of troughs, channels, raceways, and different-shaped basins through which spring water is channeled at a swift rate that enables the farm to maintain its high-density population.

The trout eggs are fertilized artificially, that is, ripe females and males are "stripped" of their eggs and sperm by a gentle massage of the abdomen. The product is dropped into a pan where it is carefully mixed. The fertilized eggs are placed in an incubator. Hatching time depends on temperatures. Under controlled temperature it takes 19 days for rainbow trout eggs to hatch.

The hatchlings are curious-looking: all black eyes and a wiggly tail attached to the yolk sack. They are kept in egg trays or concrete or aluminum troughs until the yolk is absorbed. Then they start feeding. They are fed a diet of food pellets consisting of as many as 30 to 40 ingredients, including fish meal, fats, carbohydrates, various vitamins and antibiotics. The young trout grow rapidly and are placed in progressively larger bodies of water. It takes from 7 to 14 months to reach market size.

The fish are harvested, either by draining the pond, or by dragging a seine through it (in the United States this has become a highly mechanized and speedy operation).

They are taken to the processing plant in a tank-truck. And there they are electrocuted, graded for size by a machine, passed through an "eviscerator," which can automatically clean a trout in about one second, leaving head and tail intact. Then they are inspected, prechilled, washed once more in icy water, boned, packed individually, and frozen. The whole operation, from pond to deep freeze, takes half an hour.

In spite of this amazing scientific and industrial efficiency, however, fish farming in the United States is still marginal in economic terms. The United States is only the tenth largest producer in the world, preceded by China, India, the USSR, Japan, Indonesia, the Philippines, the Province of Taiwan, Thailand, and Yugoslavia (which has developed a large inland carp, and carp-and-duck production, as well as a highly successful oyster and mussel culture in the Adriatic Sea).

In the United States, until quite recently, there has been a great deal of skepticism with regard to the potential of fish farming. Aquaculture, although scientifically and technologically feasible, has not, with a few exceptions—among them the catfish and trout industries—been a very successful cash-profit producer in the United States. Yet if the purpose is to produce food, or food and employment for large numbers of underfed and underemployed people, that is another matter.

The situation changes if one looks at the economy of the oceans not sectorially but comprehensively. If aquaculture is considered as a major source of food, if in fact it is the only way to save the world's fisheries, then why should it not be financed from the profits of other, already well-established sectors of the economy? A small "ocean development tax"—let us say, 1 percent—on offshore oil production, payable to an international organization responsible for the development of aquaculture for the benefit of all people would go a long way to get large-scale aquaculture projects going.

Vast ocean areas are deserts, the skeptics will say: blue deserts. Their waters are unproductive, and they cannot sustain a viable fishery.

Let us fertilize the blue deserts.

Enclosed Seas

There are scientists who have already thought of it. The Chinese, as we have seen, are fertilizing their huge oceanic kelp beds. Soviet scientists are working on schemes to improve the productivity of wide ocean areas by controlling the transfer of substances and energy. They are attempting to develop new varieties of single-celled algae which would utilize the sun's energy more effectively. Zooplankton, mysids, mollusks, and worms could be "planted" on a large scale. The Russians have transplanted a krill-like small plankton crustacean, named *calanipeda aquiaduleis,* from the Azov to the Aral Sea, with great success. They have introduced a number of small invertebrates, *Nereis* and *Synesnia,* from the Azov/Black Sea basin into the Caspian to feed the rapidly growing sturgeon populations.

Enclosed and semienclosed inland seas, like the Caspian, the Black and Azov Seas, the Baltic, the North Sea, the Mediterranean and the Caribbean, can be turned into fish farms. The Russians have gone a long way in this direction.

Naturally a highly productive sea, the Caspian is rich in sturgeon, Caspian white salmon, bream, and wild carp. But these are now bred in land-based hatcheries and released in massive quantities into the sea. About half of the total Caspian fish population is artificially raised.

The story of the Caspian sturgeon is exemplary. As caviar-lovers all over the world know to their cost, the sturgeon stock was sadly depleted in the 1950s. The Beluga sturgeon, in particular, is a slow-growing, mighty animal that may reach a weight of 200 kilograms in 6 or 8 years, and if it is fished faster than it grows, the species is in trouble. Additional trouble came from hydrological changes and the destruction of breeding grounds. Today, however, there are eleven flourishing sturgeon farms in the Caspian basin. Together they produce over 100 million healthy young sturgeon annually for release in the sea. About a billion have been released to date. They belong to 13 different species, some natural, some man-made like the amazing "bester." While the catch of beluga has already increased tenfold, artificial sturgeon production is still being stepped up, so that in the next few years the annual number of fish produced is expected to approach 150 million.

Meanwhile the natural reproduction of the sturgeon population has also been greatly enhanced.

The sturgeon migrates up the Volga for spawning, and then returns to the Caspian Sea. This natural sequence of events had been disrupted by the construction of the huge hydroelectric power installation at Volgograd.

The dam, over 3 kilometers long and 50 meters high, separates an upper basin from the lower course of the mighty river. It is the largest hydroelectric power works in the world.

After a brisk five-minute walk from the entrance tower, one reaches the middle of the Volga and looks down, past wooded islands, toward the distant Caspian Sea. On the right, from the ruins of Stalingrad, rises the reborn city of Volgograd with its modern factories and housing developments. Overlooking the city is the gigantic monument to the Battle of Stalingrad. The whole hillside has been terraced, and peopled with colossal statues, symbolizing the destructiveness as well as the tenacity, cooperation, and constructiveness of humankind.

And what about the sturgeon, now that their spawning grounds have been cut off by the dam?

The Soviet engineers found a rather unusual solution to the sturgeon's problem. They built an enormous fish lift, consisting of two elevators, each as large as a good-sized swimming pool, and carefully positioned in the path most traveled by the sturgeon.

Large, strong, beautiful fish, the sturgeon jump and dive, almost like dolphins, rhythmically.

Lustily splashing, the unusual passengers take their places in the elevator. One of the two elevators departs every two hours. The fish probably even know the schedule.

Slowly the elevator begins to rise. When it has reached the upper level, one side drops, the passengers are discharged and resume their migration up the mighty Volga. Tens of thousands of them make this journey every day during the spawning season.

On the return trip they do not need the elevator. They slide down through the dam's spillways.

It is not only the sturgeon, however, that has gained a new lease on life. The Caspian white salmon has been bred ashore experimentally, and about 9 million fingerlings are now released annually. In ten years' time the number will be increased to 50 million. Wild carp and bream are raised and released by the billion.

One of the most important aspects of fish production in the Caspian Sea is the importation of new species. Apart from the small invertebrates which increase the water's productivity, a number of larger fish have been transplanted. The acclimatization of the Black Sea gray mullet has been most successful. Forty to 50 thousand tons are now harvested annually. And the prospects for Pacific salmon, such as chum and silver salmon, and for grass carp and silver carp appear to be good. By 1990 the productivity of the Caspian is expected to have been increased tenfold.

Similar hopes are entertained for the Black Sea, where the Georgian branch of VNIRO—All Union Research Institute for Marine Fisheries and Oceanography—is doing pioneering work. Housed in a modest, somewhat ramshackle building at the edge of the Black Sea town of Batumi, the Institute is engaged in plankton transplants, in genetic research, and in the breeding of mullet, turbot, and Black Sea flounder. The steelhead salmon has been successfully transplanted from Oregon. Seaweeds, mussels and oysters are also cultured in the Black Sea, which gradually is being turned into a huge fish farm. Management problems could arise from the presence of a highly developed and productive offshore oil industry. At the other edge of Batumi rises a modern complex of oil refineries. But pollution control standards are severe: the uses of the sea have to be harmonized. The water purification plant, built into the refinery complex, is so efficient that the Black Sea along the shore of Batumi is clean and one can pleasurably swim right in the shadow of the refinery.

There are excellent prospects for extensive aquaculture along the western coast of the Barents Sea and some bays of the White Sea. Artificial spawning grounds for herrings have been introduced in the Barents Sea; if successful, such farming methods might be adapted to the moribund North Sea herring fisheries. The North Sea fisheries, greatly debilitated in recent years by overfishing and oil drilling, can be saved only if the North Sea is treated as one big polyculture.

Flounders are cultivated in the Barents and White Seas, and the acclimatizing of Pacific salmon (pink salmon and coho) is carried out on a large scale. The pink salmon has also adjusted to living in the Baltic.

The rainbow trout is versatile. In the Soviet Union, it is often stocked in ponds, in polyculture with carp; it is also raised in cage enclosures in the sea; and, on a large scale, rainbow trout are released into the sea. This is the work of collective fish farms on the Baltic coast.

The Russians emphasize the integration of *aquaculture and fisheries* rather than the integration of *aquaculture and agriculture*. Although the Soviet Union is well provided with streams and lakes, and carp, catfish, and buffalo fish are produced inland, the inland production is small compared to that of Asia.

"The only possible way of meeting the rapidly growing demand for animal protein," Professor Moiseev of VNIRO said, "must be a more effective and rational utilization of the potential living resources of the world ocean itself. This may become realistic only if the human attitude towards the entire problem of ocean fisheries changes radically." Mariculture, however, the Russians emphasize, is in its infancy. Although significant advances have been made and impressive results are expected in the next two decades, Soviet mariculture is still largely experimental.

The collective fish farm near Tallin in Estonia is a traditional fishing village, with a bustling harbor in which a fleet of low-powered trawlers is loaded and unloaded. (Horsepower on coastal-water fishing boats is limited, so as to prevent overfishing.) Almost everything—the boats, the nets, and the traps—is designed and made on the collective fish farm. There is also a fish processing plant which manufactures canned, smoked, and dried products.

Not all collective fish farms are like the one we visited near Tallin. But watching a community grow as it transforms an endangered and decaying natural fishery into a seafarm is an exhilarating experience.

If relatively little has been done in enclosed seas like the Mediterranean or the Caribbean, by way of fertilizing the waters, breeding native species, and introducing new ones, the reasons certainly are not scientific or technological. They are institutional and political. To turn these seas into fish farms would require new forms of cooperation among their numerous and politically and economically diversified coastal states.

The Mediterranean country most advanced in the culture of marine fish is Italy, now the eleventh-largest producer in the world. Yet Italy is exploiting only a small portion of its potential. According to recent projections, production could be increased sufficiently within twenty years to put Italy in fourth place, between the USSR and Japan. Whether such an ambitious undertaking will succeed depends on political will, economic exigencies, and institutional arrangements.

Promising beginnings have been made along the Adriatic coast. Fish culture in these parts has a long history, from the Romans through the Middle Ages when the *valli*, or large lagoons, were the property of feudal landowners. In the sixteenth century 61 *valli* were under cultivation. In the eighteenth century the cultivated area was four times its present size. Systems of exploitation ranged from the simplest to the quite complex. The primary harvests were mullet and eel, although some other species—bass and bream—were also of some importance.

The mullet spawns out at sea, during the winter months. During late winter or early spring the tiny fry drift landward, seeking the warmer and nutrient-rich brackish water of the lagoons. They feed on algae, especially blue-green algae, but also on zooplankton such as copepods, and small insects. They stay in the warm estuaries during the spring, summer, and autumn. When the shallow waters chill to temperatures below the more constant sea temperatures, the mullet returns to the sea.

The young eels, or elvers, have a much more mysterious history. For reasons of their own, the eels breed in the faraway Sargasso Sea, far below the seaweed jungle of the Bermuda Triangle. It is from here that the newly hatched larvae begin their two-year journey toward the European rivers, carried by the Gulf Stream, and using such navigational aids as the moon and the stars. Having reached the European continental shelf, the larvae metamorphose into glass eels or transparent elvers. In this new guise, they head into the estuaries, where they grow dark and lose their transparency. While the waters are warm, they grow rapidly. In the winter they dig into the mud and hibernate. During the six years it takes them to reach full maturity, they are carnivorous, feeding on a variety of invertebrates and small fish. After six years, they change into "silver eels," creatures with big eyes and silver bellies, and they are ready to return to the Sargasso Sea. Strangely, they do not eat anymore during their existence as silver eels. They consume themselves on the arduous return journey and in the final act of spawning.

The simplest form of fish culture in the valli—vallicoltura—consisted in inviting the fry's and elvers' entry but barring their exit, through a system of poles and grids of reeds, guiding the fish to a catch pond where they were caught during full moon. The average yield varied from 80 to 150 kilograms per hectare per year.

Some of the valli, however, were far more complex in their construction, consisting of an elaborate array of trenches and basins and canals and sluices and pumps, to control water levels and salinity.

The waters of the northern Adriatic play an important role in the hydrology of the Mediterranean Sea. Cooled by the cold winds blowing from the Alps, the surface water sinks to the bottom and circulates southward. The northern Adriatic, as Lord Ritchie-Calder put it, is one of the "lungs" of the Mediterranean, and this "lung" is congested: the area is one of the worst polluted, and thus endangers the health of the entire basin. Other factors, social and economic, have contributed to the decay of the valliculture during this century. Many of the valli now belong to the state, but there was no public plan, no budget, for aquaculture. Some are still in private hands, but the large landowners had neither the means, nor the knowledge, to rehabilitate the culture.

Three catastrophically cold winters, in the early 1960s, killed practically all the mullet in the valli while the number of fry migrating from the sea was getting smaller and smaller; and many of the eels in the northern lagoons were killed off by a parasite. The sickness traveled southward, wreaking havoc among the silver eels in the lagoons of Comacchio. Thanks, however, to Dr. Gino Ravagnan and his team, the commercial enterprise Sivalco and Sopal (SIRAP), and the *Centro Ittiologico Valli Venete* (CIVV), a cooperative scientific research enterprise in which 35 *valli* administrations are participating, the situation has improved.

Mullet production is limited by the availability of fry. During recent years, however, there have been breakthroughs in artificial spawning of mullets, especially in Taiwan, in Hawaii, and in Israel. In Israel, mullets have been reared in captivity, matured and, after three injections with pituitary extract, induced to spawn. The main difficulty, however, is not the spawning but the nursing of the tiny, enormously delicate, planktonic fry. When they have consumed their yolk sac—after two to three days, they simply die. Research on the nutritional requirements of baby mullets is itself still in its infancy. Thus far only one Taiwanese marine fish culture laboratory, at Tung Kang, has succeeded in rearing 500 mullets (*Mugil cephalus*) from the egg to a length of 19 centimeters. The baby mullets were kept in neon-lighted concrete tanks, in well-aerated water, and fed a mixture of zooplankton, oyster larvae, and cultured diatoms. When the technology of artificial spawning and rearing of mullet fry is transferred to the farmers on a commercial scale, "mullet could well become the most important human food product of the estuarine environment." (John Bardach, *Aquaculture*.)

Meanwhile, within the limits set by the availability of elvers caught from the wild, eel production in the *valli* has increased tremendously. While, with the traditional, extensive method of cultivating in the Comacchio *valli*, production was about 20 kilograms per hectare per year, the new, intensive methods, in hothouses, following the Japanese example, have raised production to 4 tons. Growing time, in these hothouses, furthermore, has been reduced from six to two years.

Two favorite Mediterranean fishes, the *branzino* or *spigola*—a kind of bass—and the *orata*—a kind of bream—were becoming scarce and expensive. Now the bass is being farmed successfully in the *valli*, and, it seems, a breakthrough in the farming of the *orata* is imminent.

The total brackish-water area of the Mediterranean where indigenous fish could be cultured is well over 100 million hectares. If, however, one were to apply the Russian concept and consider the whole Mediterranean as a big fish farm whose waters are to be fertilized, whose native fish are to be land-produced and then released, while new species could be transplanted and acclimatized, then the potential would be enormous. If Black Sea productivity can be increased tenfold, why not apply the same methods to the Mediterranean, or the Caribbean?

China: Fish being loaded in live-fish boats for shipping to Hong Kong.

A breeder carp on a
commune near Shanghai.

Fish traps and pond walls, Panay, the Philippines.

Another enclosed sea which lends itself to seafarming is Japan's Inland Sea.

Although seriously polluted, the Inland Sea is becoming—*must become*—a seafarm, if Japan is to adapt to the new Law of the Sea and its restrictions on distant-water fishing. Besides the seaweed, oyster, and shrimp developments described in earlier chapters of this book, the Seto Inland Sea provides great scope for fish ranching. There the "king of the fish," the sea bream, is sent to pasture. A member of the Sparidae family, the sea bream is a fascinating animal. Its origin has been traced back to Cretaceous times when it flourished in the Tethys Sea, which disappeared after India split off and moved away from what is now Australia and Antarctica. Since then, the bream—six subfamilies of it—has made its home in the Atlantic, Pacific, and Indian Oceans as well as in the Mediterranean and other warmer seas. The bream leads a rich and varied life: in its youth, until it reaches a length of about 10 centimeters, it is a sexless being; then, though physiologically hermaphroditic, the bream is functionally a male. Its female organs mature somewhat later. When the fish has reached a length of 20 to 25 centimeters, it becomes a fully functional female. She may grow up to a length of 80 centimeters, a weight of 8.5 kilograms, and may reach the venerable age of 25 years. The bream is the queen of fish rather than the king.

The Japanese have a particular fondness for the bream, especially the red sea bream, *ma-dai*. It is a good-luck fish, served with plum blossom on birthdays and weddings.

Between the 1930s and the 1960s the sea bream was threatened with extinction; only recently was it rescued by the science of farming. In captivity, spawning may have to be encouraged by hormone injections. The ripe fish are removed from their floating pens and transferred to spawning tanks ashore. The fertilized eggs are removed from an outlet at the center of the tank with a fine-meshed net and placed in tanks with running water. After fifty hours they hatch: strange creatures, their mouths and anuses still closed until the egg-yolk is resorbed, which takes about 96 hours. After that they open their mouths and start feeding.

Japanese farmers have reconstructed a complete ecosystem within their labs to duplicate that of the Inland Sea. Thus the alga *Chlorella* is reared in seawater tanks, fertilized with ammonium sulphate and calcium phosphate. The small rotifer *Brachionus* is cultured on the *Chlorella*. It multiplies rapidly, which is just as well because each sea bream fry eats from 200 to 500 of these little zooplankton animals every day. Thus feeding a crowd of 15,000 fry in a rearing tank requires a daily supply of 5 to 10 tons of *Brachionus*. Copepods, which are to be found among the rocks of the seashore, also feed on the *Chlorella* and are

Yellow tail and seabream circular cage, Japan.

added to the diet. The tiny brine shrimp Artemia, and the larvae of oyster, acorn barnacle, and blue mussel are carefully prepared. Vitamins, wheat protein (gluten), and minerals are added. When the young bream is about a month and a half old—it is an appealing little fish with big round eyes and lips that seem ready to speak—the diet is further enriched with minced shrimp and other fish. To enhance the appetite (and the growth rate) of the baby bream, it is kept indoors in heated tanks.

Most fry that have reached a length of about 4 to 5 centimeters are taken out to sea and released.

Cage Cultures

Many, however, are sold to farmers, who raise them in floating net cages. The cages have frames made of steel piping or bamboo or cedar rods. They are floated by styrofoam cylinders. The recently invented submersible cage can be lowered on ropes below the turbulence level and thus is less exposed to winds and waves. This has extended the seafarming area into the open sea, away from congested and polluted coastal areas.

Some adult bream are reared in cages aboard a specially built ship, the *Carina*, which resembles a floating dock. It is the property of the Seto Inland Sea Fishery Association and is equipped with four rearing pens of wire netting, a laboratory, and windows in the hull through which the fish in the pens can be observed. When the Inland Sea gets cold in winter time, and the bream stop feeding, the *Carina* takes them for a winter vacation in warmer waters.

Another fish, frequently farmed together with the bream and in the same type of net cage, but on a much larger scale, is the yellowtail. In quantitative terms, the yellowtail culture is in fact by far the most important of the open-sea cultures in Japan, having by now completely displaced the capture fishery of the adult of this species.

Although artificial fry production has been achieved at the experimental level, the technology has not yet been commercialized. Thus the yellowtail is not yet fully domesticated, and fry have to be caught in the wild when, during early summer, they migrate to the coast of Japan,

A Cambodian cage-boat community.

Sea bass in cages being fed, Songkhla, Thailand.

riding the Kurishio current. They usually are found playing around some floating seaweed, and are caught by encircling nets. Yellowtails are fed on sand eels, horse mackerel, saury, and anchovy. This is a rather costly diet. Food, in fact, is the highest item on the budget of the yellowtail farmer, as of the cage farmer in general. Research is now being done on synthetic commercial feed for yellowtails.

Keeping fish in cages is rather like keeping cows in stables. It is surprising how fishes accustomed to roaming the world's oceans from one end to the other, fishes as muscular and as swift as racehorses, adapt to a stable existence. One might think that they would fret, and pine, and even die. But they don't. Sociable and cooperative, they grow faster and fatter in captivity than in the wild.

Cage culture originated in Cambodia many centuries ago. The Cambodians built shipshaped cages; indeed, whole ships were built as a floating cage, with a house for the fisherman and his family built on top. In this type of vessel the fisherman transported his catch—snakehead, catfish, and sand goby fry—from the Great Lakes area to the markets in the capital city of Phnom Penh. During the long voyage the fish were fattened on kitchen offal. Fungus infestations were prevented by dump-

Cage culture in Hong Kong.

A farmer with a fry trap.

ing cow dung and the bark of certain trees, or a mixture of salt, mud, and leaves, into the cage. Modern science has yet to figure out how this folk medicine works! If the journey began in February, by May, when the fish arrived in the city they had grown to market size. Farming, transportation, preservation were all taken care of at once.

Cambodian emigrants introduced the technique to southern Vietnam. They are still there, living together in floating villages of their own.

From the Mekong River and its tributaries, the practice of raising fish in cages has spread to Thailand, Indonesia, Japan, and Hong Kong.

In the Uthai Thani Province of Thailand, everyone has a cage under his home. One gets up the lotus-choked river in a boat with an outboard motor which can be lifted clear of the water when the tangled weeds are too strong to cut through. On both sides of the waterway, there are houses, modest huts, with landings and front porches; and under the porches are the cages. One lifts a couple of planks, and there they come, swimming about, waiting to be fed: catfish, tilapia, and gourami. Their daily diet: 50 kilograms of cracked rice, and 5 to 8 baskets (15 kilograms each) of trashfish, which the owner fishes himself; for he is a fisherman. There are also some greens ready, and vegetable tops, for the gourami is a vegetarian. In the back of the house is a small nursery cage for the fry, which will be moved up front next year. And on the porch, on top of the cages, there is the "processing plant"—an array of tables and mats for sun-drying the fish, and a place for a fire for smoking it. The fillets of smoked catfish, on long skewers, are a delicacy and fetch a high price on the market—as much as $7.50 per kilogram!

Rows of rafts lie in the sheltered Hong Kong bays. Here grouper, sea bream, red snapper, and pride fish are raised. There are six cages to a raft. And a good-sized farm may manage eight rafts. Here, too, there are little huts on the top of the cages, where the farmer and his wife may sleep. To help ward off poachers there are lean, lusty dogs of no particular breed, which make a fearful barking chorus once one of them gets started. The poacher has no chance. (In Thailand, the catfish ponds are watched by geese: geese which make a great row when anyone gets near, and which also fertilize the ponds.)

Lately cage culture has been introduced in the United States and in various parts of Europe.

Salmon and rainbow trout cage farming, although only recently introduced in the fiords of Norway, appears to be a great commercial success. It is anticipated that the quantity of cultured salmon will, in the foreseeable future, surpass that of captured salmon, a part of which, in any case, is itself cultured and released. The cage culture of salmon from egg to market size is an expensive undertaking. On one Norwegian salmon and trout farm, the salmon in four large pens consume a daily supply of 1,000 to 1,800 kilograms of fresh fish. Thinking in terms of world hunger, that seems quite a waste. Economically, however, the undertaking is worthwhile. In this rich part of the world, if the fish were not fed to the salmon, they would be ground into fish meal and fed to hogs and chickens—and these fetch a lower price on the market.

V. R. Pantulu of the United Nations says that cage culture "can be considered to represent a very advanced type of aquaculture; its productivity is 10 to 20 times higher than that of pond culture for comparable inputs and area."

Mitsuo Iwashita, a Japanese expert, recently designed a floating hatchery. It can be installed on any old supertanker, taken out of mothballs. Outdoing the *Carina*, it is equipped with a spawning tank, a rearing tank for larvae, a tank for the mass production of food plankton, a rearing tank for advanced fry, and a larger rearing tank where young fish are kept until they can safely be released. The ship is also equipped with a laboratory; pumps; a computer programmed to regulate water temperatures, salinities, and oxygen content; a tie-up with a satellite; and radar to locate schools of fish, and to catch breeders and fry. As the tanker cruises along it picks up tuna or whatever high-seas species might be desirable. It accommodates the life cycle of the fish on board, and releases their offspring by the hundreds of thousands. Mr. Iwashita is convinced that "the Pacific Ocean could become a large aquarium."

12

"WHERE THERE IS WATER THERE SHOULD BE FISH"

One country alone, China, produces almost half of the total amount of fish farmed in the world. The Chinese grow over 2 million tons of fish every year. One and a half million people are employed in fish farming. While aquaculture has deep roots in Chinese history, production has multiplied by leaps and bounds over the past two decades.

Although statistical figures are not easy to obtain in China, the best available sources indicate that production increased from a mere 370,000 tons in the 1960s and will probably exceed 3,000,000 tons in the early 1980s.

How has this prodigious development come about? What can the international community learn from it in the struggle against world hunger and unemployment?

Obviously, some aspects of the achievement are peculiarly Chinese, rooted in natural, historical, and social conditions, and it would be futile to try to export them. Other aspects, however, can be "universalized." They may embody truths whose time has come. They could constitute a major Chinese contribution to development planning in the 1980s, within the framework of a truly new international economic order.

The success of Chinese fish farming is based on a few interacting principles; many countries recognize them, but nowhere else are they integrated in their application.

The first principle is the integration of aquaculture and agriculture.

Aquaculture and agriculture have a common matrix: *water conservancy and management*. Water levels must be maintained to forestall floods and droughts and to supply irrigation systems. This requires the building of dams, dikes, irrigation and drainage canals, lakes, reservoirs, and ponds. At the time of Confucius, irrigation systems covered millions of square miles. An imperial edict of the year 111 B.C. states:

Agriculture is the basis of the whole world. Springs and rivers, irrigation ditches and reservoirs make possible the cultivation of the five grains. In the empire there are innumerable mountains and rivers, but the small people do not understand their proper use. Hence the Government must cut canals and ditches and build dikes and reservoirs to prevent droughts.

So advanced was hydraulic engineering that in 95 B.C. serious thought was given to diverting the entire Yellow River so as to cut off its vast northern excursion into the Gobi Desert area.

These ancient hydraulic works are a testament to the Chinese genius for mobilizing huge masses of people. A text from the late third or early second century B.C. prescribes:

I request that you establish water conservancy offices in each district and staff them with men who are experienced in the ways of water. There should be one high official and one deputy, with just enough labor corps brigadiers, section commanders and administrative assistants, to meet the need. Then for the area on both sides of each river select one man as chief hydraulic engineer. Order all these to inspect the waterways, the walls of cities and their suburbs, the dikes and rivers, canals and pools, government buildings and cottages; and to supply those who are to carry out the repair work in the districts with just enough men.

There are texts, dating from the first century B.C., that manifest an *interdisciplinary* approach to hydraulic planning, associating pure scientists, mathematicians, with the technicians and administrative officials to ensure the success of a project. All this has an incredibly modern ring.

The Chinese required their hydraulic works to serve many purposes: river control, navigation, irrigation, agriculture, and the culture of fish. That, from the beginning, they served this last purpose as well is indicated by a very ancient representation of irrigated fields, in a Han tomb model of black pottery from Phengshan (Chekiang Provincial Museum at Hangchow). The reservoir, from which channels flow to four fields where rice is grown (indicated by piles of rice straw), is characterized by two fishes carved in the clay.

The ruins of dikes, canals, and reservoirs are eloquent testimonies of a great past. And if the present leadership listens to the people, it also listens to the collective wisdom of the past. Thus in the past three decades, working mostly with their hands, the Chinese have rebuilt a system of water conservancy and management. Beginning in 1950, they have built 130,000 kilometers of dikes and dams and an irrigation network compared to which the famous Chinese Wall, snaking along 6,000 kilometers of mountain ridges, is a minor achievement.

Hydropower has promoted the mechanization of farming and the wide replacement of animal or human-driven pumps with electric pumps.

The construction of canals, connecting main rivers or lakes, has also benefited navigation, which had to be included in the multipurpose water management planning.

The Chinese have a saying, "where there is water, there should be fish." Thus their ponds, rice paddies, rivers, canals, lakes, and reservoirs have been stocked with billions and trillions of fish. Whereas in 1955 the area available for fish farming was no more than 312,000 hectares, it is now over 10,000,000 hectares.

While water management is the common matrix of agriculture and aquaculture, these two still interact in Chinese thinking in numerous other ways.

The Chinese use no inorganic fertilizer in ponds, lakes, or reservoirs. Land farm animals provide the organic manures that fertilize both croplands and ponds. The Chinese say it takes 15 pigs to fertilize one hectare of water. In China's diversified economy the farming of land animals complements the farming of fish.

The grass and the vegetables that are grown between the ponds are fed to the fish—grass carp. The wastes of the grass carp, in turn, provide food for other fish. The Chinese say, "feed one grass carp well, and you feed three other fishes."

Wastes not consumed by other fishes accumulate at the bottom of the pond and form a fertile humus. The pond humus is collected four or five times a year and is used to fertilize cropland. Everything is recycled and reused.

There is a strong connection between the growing of rice and fish culture. Thus in Kwangtung Province, both rice and fish are cultivated in a paddy field by a work brigade. A work brigade consists of about 1,000 men and women managing tractors, trucks, large boats, and other heavy equipment. The paddy field is just one among many other activities of the brigade.

A part of the paddy—about half an acre—is given over to intensive fish culture. It is fertilized with pig and cow manure every day.

A week after the rice is transplanted, the paddy is stocked with 9,000 carp and tilapia fry. Fed daily with green grass, they are kept there until harvest.

The fish help the rice crop in various ways. They eat the insects, such as leafhoppers, that attack the rice plants; they stir up the mud, thereby increasing the nutrient intake of the rice plants; and they enrich the water and nourish the rice plants with their manure.

The first artificial spawnings of pond-reared Chinese carp were achieved in 1958. But whereas, in other countries, artificial fry production is the virtual monopoly of a few government or university stations from which the fish are delivered to the farmer, in China a massive transfer of technology has given every commune its own fish-breeding station, and, in a way, every farmer has become a scientist. The Chinese method of integrating science, politics, and economics is called "three-in-one." Decisions at every level are made by teams of scientific researchers, workers, and political administrators. Research needs are determined at the farm or factory level, where much of the research itself is carried out. There are five major research institutes, open to farmers—experience of farm or factory work is a prerequisite for higher studies—and researchers in turn take time out helping on the farms.

Practically every farm has a brood pond, a spawning pool, a hatching pool, and a nursery pond. The polyculture of the pond system fills every "ecological niche" with different kinds of carp: the grass carp feeds on the large surface vegetation; the silver carp, a mid-water dweller, eats phytoplankton, while the bighead eats zooplankton; at the bottom are the common carp and the mud carp, both detritus-feeders, and the black carp, which feeds on snails, and in some regions, another bottom dweller, the bream.

In China the carp is called "family fish" or "home fish." They are stocked in various proportions, according to ecological conditions. The grass carp and bighead breeders are imposing-looking fishes, sometimes weighing more than 12 kilograms.

Experiments in crossbreeding different kinds of carp have been done in China. The mirror carp and the common carp, when crossbred, have produced a hybrid which grows much faster than either of its parents. The breeders reach sexual maturity in the ponds in 2 to 5 years, and continue to breed for about ten years, during which time they develop distinctive personalities. They come when called. They become "family fish" while at the same time producing their own huge families.

When they first mature, they are netted, left in the crowded net for a while, and then released. This process is repeated several times, so as to accustom the fish to handling. Thus trained, they are taken out and given their injections. The injected fish—two males to one female—are placed in the spawning pool.

The fry feed on the natural plankton growth, stimulated by the fertilization of the pond, but in addition they are given home-prepared "fine foods": wheat flour, fine rice bran, egg yolk, soy bean and peanut powder cake. The Chinese never use commercially prepared food for the fish.

Digging a pond on a communal fish farm, near Shanghai. Two thousand workers dig ponds and irrigation ponds full-time. Occasionally another 8,000 workers are employed. Annually, they move an average of 3.5 million cubic meters of earth equal to the size of the Boulder Dam. Slender girls carry as much as 150 pounds.

Seining of a breeder pond.

The feeding with "fine foods" is continued for some time after the fish are transferred to the polycultures of the larger rearing ponds. The grass carp, however, are now started on "coarse foods," fresh-cut grass, which at first is chopped up very fine.

In production ponds only "coarse foods" are served: a large variety of fresh-cut grasses, elephant grass, vegetable tops, and water hyacinths. Pigs, penned on the banks of the pond, provide organic fertilizer to stimulate the production of natural food for the other species.

Ponds in China are laid out in such a way as to catch as much wind as possible. The oxygenation of a large and intensively cultivated pond, however, is provided by an aerator, a floating power-driven rotary device, made in commune factories. This machine, called a "yield-increasing machine," not only oxygenates; the stirring up of the water also circulates nutrients, stimulates plankton growth, increases the feeding activity of the fish and raises their metabolism. Where yield-increasing machines have been installed, ponds have increased their yield about 30 percent.

On some farms and communes, fish farming is the primary activity, and the raising of livestock and the cultivation of vegetables and rice take second place. On other farms, agriculture—for example, the culture of soybeans, sugarcane, mulberries—is the main activity and fish farming is secondary. On the latter, the agricultural land areas between ponds are greater. On both types of farm, pond production is intensive.

Chicken, cattle, pigs, sheep, ducks, and geese are raised on the shores of China's many lakes and reservoirs. They fertilize these large bodies of water in which fish culture, however, is far less intensive than in the smaller ponds.

One of the maxims of Chinese fishing, to be found on posters and in manuals, is "Lay stress on culture and simultaneous development of culture and capture." Another reads: "In fisheries, do stocking and catching in rotation." A great deal of inventiveness has been lavished on the perfection of harvesting or catching techniques, with capture being conceived as a phase of culture.

Inventiveness is encouraged, and innovations are often made, at the level of the commune, by the workers and farmers themselves. One such invention comes from the Ho Law Commune, Wu-hsi (in the Shanghai region). The large seine net, dragged by men in boats, is floated, in the center portion, by truck tires on which wooden frames are mounted. The upper rim of the seine net is hooked to these frames, and thus a fence, over half a meter high, prevents the fish from jumping back into the pond.

In some large lakes, fish are driven towards one corner before being

Hatching and nursing tanks.

A hatching tank. Water is piped in from below. A water spout is equipped with a flange, which keeps water in fairly fast circular motion, to simulate river conditions and keep eggs separate.

caught. Up to fifty fishermen may give chase in their boats. They drive the fish before them by beating the water with their oars and shouting. When the fish have been driven a certain distance, a net is set behind them, to block their retreat. Then, after more beating and shouting, a second net is set. The first net is eventually placed in front of the second, and so on, in an intricate pattern throughout the lake. It looks like the giant version of an oriental boardgame. Sometimes it takes as long as 20 days to capture the fish. The seine nets used for lake fishing may be three miles long.

When the fish have lost the game—the game is designed in such a way that the fish must lose—they are driven into a trap, which is incorporated in the last blocking net.

The whole system is carefully depicted on a poster at East Lake in Hupeh Province (there is a special East Lake Breeding farm to provide the fingerlings for this large lake).

Posters and manuals feature an array of seines, bag nets, cast nets, dragnets, set nets, and traps of all sorts. The idea behind the graphic instructions for their manufacture and use is that anyone can manufacture and use them. The adoption of the new techniques has transformed the life of the fisherman: he has changed from poor lake-dweller to well-to-do land-based member of the agricultural communes.

China's progress has been spectacular. It should also be observed that when Chinese technologies and methods have been applied experimentally in other developing countries, encouraging results have been achieved.

There was a poor fishing village, San Francisco, on the Pacific coast of Mexico. Despite the great beauty of its bays, jungles, and mountains, the living conditions of the village's one thousand inhabitants were miserable. Infant mortality was high, and undernourishment kept energies at a low level. A few straggly chickens, a haggard horse or cow constituted its agriculture. In good seasons the villagers would go out and catch 200 kilograms of fish—and most of it had to go to the market in Vallarta. Without a surplus it was not possible to install electricity or other public or hygienic services. Besides a few wilted greens, the wretched village store offered only canned goods, which stayed on the shelves because no one could afford them.

The Center for Economic and Social Studies of the Third World in Mexico City recently initiated an experimental fisheries project in this village. It was directed by an enterprising Japanese who had traveled extensively in China and brought with him the Chinese posters and manuals which could easily be adapted to Spanish use. A set net, on the Chinese model, was made to order in nearby Vallarta. A bright yellow "guide net" leading 300 meters out from shore, diverts the fish into the deeper water where they are funneled into a large sack. The sack is 75 meters long, 25 meters wide, and 13 meters deep. The entire contraption is anchored with 3,000 sandbags, floated with hundreds of local glass bottles, and marked with banderoles and lights to avoid encroachment by boats. The meshes of the net are wide enough to allow young fish to escape. Large numbers of the captured fish can stay in the net for weeks at a time, a necessity in a community lacking refrigeration.

The cost of this equipment, about $1,200, was advanced by the company that built it, and repaid by the village within the first year of use. Instead of catching kilograms a day, they now caught tons: the villagers had enough for their own consumption and for wholesaling in the market in Vallarta.

One third of the revenue from the catch goes to the fishermen, who are now making $100 per family per month; one third goes to the Center for the costs and overhead of the project; while one third is invested in other local projects. Thus two more nets have been constructed, and the daily catch, in good seasons, may now total 20 tons. Glass-fiber boats with outboard motors, suitable for operating the set nets, have been constructed; a refrigeration plant and a fish meal factory have been built in the village (by independent enterprises); a workshop has been set up to salt, smoke, and dry fish and utilize leftovers in fish croquettes and sausages; and there are courses in which more than two hundred local students learn from technicians and fishermen about meteorology, oceanography, the behavior of different species, and the resistance of different materials exposed to sea and sun.

The Chinese posters and pages of the manuals are exhibited on the walls of San Francisco's new fisheries museum. Its walls are of bamboo, the traditional high conical roof is made of palm straw. The floors consist of sections of local tree trunks cemented together. Showcases are made of local coral rock, internal divisions of fishing nets. Exhibited in the museum are all the local species of fish—about 50 of them—and all the tools and nets and traps and boats. The design, proportions, textures, and setting make this easily one of the most attractive museums in the world.

San Francisco's standard of living has improved enormously. There is now electric light; there is a community nurse handling childbirths, emergencies, and preventive medicine such as tetanus shots; and three big rooms containing an impressive array of household goods, groceries, and clothing have been added on to the local store.

The unpolluted warm waters and coastal embayments could sustain exemplary cultures of fish, mollusks, and crustaceans. However, the San Franciscans have yet to apply the Chinese precept: "Lay stress on culture and simultaneous development of culture and capture." They remain at the mercy of seasonal changes and they are in danger of overexploiting what may first appear to be an inexhaustible resource.

In China, fish is the cheapest source of animal protein. A kilogram of fish on the market costs 30 to 40 cents while the cheapest cut of pork costs about twice that much, and duck and chicken may cost up to $1.50 per kilogram.

The price of fish has been kept as low as it is because processing and transport costs are kept to a minimum or eliminated altogether; the fish are sold fresh, mostly alive; communes eat their own fish, and production in or near cities—for example, in the lake in Peking—is intense.

Until recently the people of some inland regions were not used to eating fish, but they quickly caught on when they saw their mountain reservoirs stocked with fingerlings and the bounty of the catches, and the feasts of the working brigades.

People's eating habits are perhaps the most conservative of all habits. But even they can be changed. A Chinese sage of the fourth century B.C. said, "Changing to the utmost is the means by which to respond to

The end product of a fish farm south of Canton.

things; only he who holds fast to the conception of the unity of the universe can do this."

If we hold fast to the conception of the unity of the universe we may yet achieve the changes which are necessary if all people are to be fed.

The oceans are the rivers and lakes of the world community. We can learn from the Chinese experience to make ocean conservancy and management the basis for international development. This unified approach includes agriculture, animal husbandry, hydroelectric power generation, navigation, aquaculture and, going beyond the Chinese experience, mineral production. By the end of the century, we might expect to achieve similar results—including an enormous increase in husbanded marine fish.

We must learn to "lay stress on culture and simultaneous development of culture and capture." We must learn to follow the maxim "Where there is water there should be fish."

The fish hospital, Toba Bay, Japan.

IV
ON THE
HORIZON

13

CONSTRAINTS

Seafarming is an activity in which we can repose much of our hope for a better world. Seafarming can help us solve one of the most pressing problems in the world today—hunger.

Projections, made by international organizations like FAO, or national government agencies or scientific institutions, may be right, or they may be wrong. Projections often are dismally wrong, because the data on which they are based are too complex, not quantifiable, or quite unpredictable.

As we have seen, however, the physical or natural potential of aquaculture is, for all practical purposes, boundless. The basic element of our planet is water. All life derives from and depends on it. Water covers three-fourths of the planet and envelops and penetrates our landmasses. To think in terms of land use and management only is too confining. If we consider water to be outside our system, we are going to ruin it. By ruining the water, however, we ruin the land and we ruin ourselves. It is by means of water conservancy and management that we shall enlarge the physical scope of the world economy, and it is only by these means that both aquatic and terrestrial sectors stand to gain. This is what we must learn. This is the Age of Aquarius.

The scientific and technological potential, likewise, is boundless. We are always just at the beginning of knowledge, and the more we know the more we know how little we know. Science and the mystery of the unknown stand in a relationship of direct, not indirect proportion. They wax and wane concurrently.

But even the very little that we know about the ecology of our living resources, about genetic options, about nutritional requirements and preventive and curative marine-veterinary medicine, if applied systematically, could increase world production a hundredfold over the next fifty years. And the frontiers of science and technology are rapidly being extended. So that what looks like science fiction today may be commonplace fifty years from now.

Thus the constraints on the development of aquaculture are neither natural and physical nor scientific and technological.

When we come to the economic, social, and political aspects, the picture changes. Economic, social, political and legal constraints on aquaculture take different forms in different parts of the world, but they exist everywhere.

In most Western countries, aquaculture is neither fish nor fowl. That is, it is not considered part of the fisheries or part of agriculture. Thus there is no governmental infrastructure; property rights encroach; loans and subsidies are difficult or impossible to obtain; fish farmers' rights are unprotected, and an application for a permit to start an aquaculture enterprise may be the beginning of a Kafkaesque trial.

In Hawaii recently, Dr. Fujimura's Institute made a survey of aquaculture financing. The Institute sent a letter to a hundred lending institutions, asking how many aquaculture projects each one was financing. Seventy-five of the hundred lending institutions replied, "aqua-what?"

Aquaculture is thought to be an exotic mystery, frequently confused with acupuncture.

To initiate a particular aquaculture project in the state of Hawaii required 33 permits, which were only obtained after a period of 18 months, and upon payments totaling $10,000.

In 1970, a pilot project on the Development of a Commercial Pacific Salmon Cultivation Business was initiated by Domsea Farm Inc. and the National Fisheries Service in Washington. "The concept of salmon farming was new," J.M. Lindbergh reported to the Aquaculture Conference in Kyoto (1976), "and an act of State legislature was required before commercial salmon farming became legal. Thereafter a license had to be secured from the Washington Department of Fisheries and a lease had to be negotiated with the Washington State Department of Natural Resources. Before any floating facilities were installed, a permit was required by the U.S. Army Corps of Engineers; and before the Engineers would issue a permit, they had to advertise the application publicly and also receive positive clearance from the U.S. Navy, U.S. Department of Commerce, U.S. Environmental Protection Agency, U.S. Department of the Interior, Washington Department of Fisheries, Washington Department of Ecology, Washington Department of Natural Resources, Washington Parks Department, Washington Department of Health, and, finally, the Kitsap County Shorelines' Management Group. While most of the agencies had a legitimate reason to be involved, many appeared to have overlapping requirements. A considerable amount of time and paperwork was involved with these negotiations before final clearance to proceed with the operation was obtained."

It takes a true pioneer to hack his way through this sort of legal jungle.

Problems of ownership frequently get in the way of aquaculture developments. An American entrepreneur recently embarked on a survey of 5,000 miles of coastline to locate suitable sites for a shrimp farm.

"Since the best type of marshland for this purpose is the least attractive for other forms of commercial development, he felt sure he could obtain the necessary property cheaply," Bardach notes. The entrepreneur guessed wrong: "Conservation laws and restrictions on tideland leasing made it virtually impossible to lease acreage suitable for shrimp culture in any of the eight States where it would be feasible."

The private use of waters in the public domain, such as great ponds, public reservoirs, lakes, rivers, and the littoral of the seas also causes problems. The animals in these waters are customarily considered part of the commonwealth, and any one is free to fish. Such a custom is hard to change, and it has hindered the growth of aquaculture in many parts of the world. Further it is claimed that: farming enterprises in such waters block navigation by floating pen structures; they are unsightly; they interfere with commercial or sport fishing; and they cause organic pollution.

Fish farming has not been defined in legal terms. Fish cages, for example, are not classified as fishing gear. As a matter of fact, they are not classified at all. Thus, while commercial salmon fishermen are allowed to shoot seals that interfere with their nets, salmon farmers using cages are not allowed to protect their stocks from marauding seals. (Needless to say, this is not meant as a plea for a permit to shoot seals!)

Living in a legal vacuum, the fish farmer has great difficulty in obtaining subsidies and loans. In Great Britain, "the fishfarmer does not appear eligible for capital grants available on farm buildings or development grants in the region. Fish farming is neither trade nor industry nor, for that matter, fishing." (Tony Loftes, "The Fish Farming Lobby," *The New Scientist*)

Neither is this legal vacuum contained within the boundaries of national jurisdictions. Thus sea ranching causes a number of international problems. For what happens if fish, nursed and released by one country, are fished by another? In international law there is no such thing as a national right of property in a herd or body of wild animals. As these herds are domesticated and technologies for marking and feeding them develop, existing international law will prove inadequate. It is curious that while from Blackstone on the legal relationship between man and the beasts of the land has been defined at great length, the relationship between man and the animals of the sea has not.

In recent years the FAO has organized a series of regional workshops to advance the development of aquaculture in areas where it is lagging. The Africa workshop reported that "specific laws for aquaculture do not exist in most countries in the region. . . . Legally speaking, aquaculture is neither agriculture, animal husbandry nor an industry, and so none of the laws relevant to these are applicable to aquaculture." Fishery Acts, the FAO reports, may be partially used to protect aquaculture but, in general, they are shaped by the requirements of common-property resources, and are insufficient for the purpose of an industrial aquaculture. A code of laws is needed: to enable the acquisition of suitable sites and the construction of appropriate facilities; to provide incentives normally given to new food production industries, such as subsidies, grants, loans, and tax holidays, water and power supplies at special rates; to import essential equipment and supplies; to protect from poaching and destruction of stock; to control fish and shellfish diseases; and to market aquaculture products. All these measures would greatly encourage investment in aquaculture. The lack of investment capital is one of the major constraints on development.

The biggest constraint, however may well be human psychology. Human nature can comprehend profound evolutionary changes after they have occurred, in history books, or as myths. While they are going on, the human mind tends to shut them out. The "inner limits" are the real constraint on future development.

It is not unusual for the transfer of technology, which is essential for the growth of aquaculture, to be considered as an economic problem that can be solved with money. Few researchers have begun to look into the limiting factors of psychology. Discussions at a recent seminar-workshop on artisan fisheries development and aquaculture in Central America and Panama revealed that, for example, new boat forms should not be imposed from outside—"that they should be developed with the cooperation of artisan fishermen. The vessel technician should work with the fishermen, trying new designs and asking the fishermen what they want. This does not necessarily mean that the technical expert always does what the fishermen suggest. What is important, however, is that the fishermen are involved from the outset, and they are more likely to see the innovation as fulfilling their needs. In fact, it will be more likely to fulfill their needs." The way in which new technologies are introduced, and by whom, is a factor of success at least as powerful as money. New forms of cooperation—for example, the formation of a cooperative to take over marketing—may be rejected if they are perceived as threatening the vested interests of the middleman, and he has sufficient power and prestige to wreck the project. The Central American workshop reminds us that "Any change which results in a differential distribution of power, wealth, or prestige, is bound to stimulate resistance within a community."

The expansion of aquaculture may bring to the sea people who are accustomed to the ways of the land. This may cause problems. "A change of occupation from a land-based occupation to marine fishing," again to quote the Central American workshop, "is a change of completely different magnitude....Preliminary work has suggested that marine fishermen are more active, aggressive, courageous, independent, and future-oriented than individuals with land-based occupations, such as factory workers and farmers."

It has often been noted by students of society, from Havelock Ellis to Herbert Marcuse, that urban industrial civilization tends to create "passive" individuals, fragmented, "one-dimensional" beings. A culture focusing on aquatic resources, on the other hand, is an integrated polyculture. As the principles of integrated management of the multiple uses of water at the international, national, and subnational levels are reflected, so to speak, in the waters of each polyculture, it is not too much to hope that these same principles will be reflected in the communities which run the polycultures, in the countries which shelter these communities, and in the world community at large. Aquaculture, understood as *culture*, may tend to produce a more rounded, better balanced human being, capable of taking initiatives and of making decisions.

14

ACTION

As the importance of aquaculture for development in general is becoming more obvious, more people are calling for action: action at the national and international level, East and West, in developing as well as in developed countries.

Take a look at one of the poorest countries, Bangladesh.

The Asian Development Bank in 1978 approved a concessional loan of $18 million to Bangladesh for a nationwide aquaculture development project. The project, to be completed over 5 years, is expected to increase fish production for domestic consumption by about 40,000 metric tons a year, and to generate foreign exchange earnings of over 6 million dollars annually from shrimp exports.

The project consists of 8 major components:

—construction of 5 fish hatcheries for fish stocking in impounded inland waters;

—establishment of a shrimp fry hatchery, with some 300 hectares of shrimp farming ponds;

—construction and establishment of 100 hectares of fish pens at three different sites in inland waters;

—construction and placement of 1,600 fish cages covering a total of 400 hectares of Lake Kaptai;

—some 120 family-size fish/shrimp polyculture ponds, each of 4.5 hectares, along with nursery and demonstration ponds, an ice plant, offices, and housing facilities;

—a plant to manufacture 240,000 pounds of gill fishing nets a year;

—fish processing facilities, at 4 sites, introducing new and simple technology for drying fish and shrimps, using polyethylene tents;

—three small ice plants and storage facilities, each able to produce 10 tons of ice and store 50 tons of fish per day.

About 2,000 fish farmers will directly benefit from the fish pens, cages, and ponds provided under the project, while an estimated 100,000 rural families will enjoy increased incomes as a result of stocking fish in impounded waters.

The project will also provide an average of more than 6,000 full-time jobs annually for the 5-year construction period.

A sum that would not suffice to finance one-third of one day of the arms race of our dying military/industrial culture can thus be life-giving, in the aquaculture of the future.

Take a look at one of the richest countries, Japan.

Japan is using computer-controlled fishfarms to compensate for the 600,000 tons of fish a year the country is losing as a result of the new Law of the Sea. Starting in 1978 the farms, occupying whole bays and inlets around the Japanese coast, are expected to produce up to 500,000 tons of fish (one-fourth of the Chinese production) within ten years, with an adequate growth rate after that. Production will be increased by controlling each stage of cultivation. Fish fry are produced in hatcheries, and new nutritional studies are in course to accelerate fish growth. New methods are being developed to keep the fish from straying too far. Sound screens may be used at an early stage. Monitoring buoys will control water temperatures, which can be raised by effluents from power plants. A computer will control the timing and the amount of the catches. Obviously such a program must be subsidized by the government. It seems almost unbelievable that all that it cost the government, however, was $200,000 for the first year.

Take a look at the war-ravaged region of the Mekong Delta.

Its greatest natural resource, the focus of reconstruction, is water from one of the greatest rivers of Asia, with its waterfalls, lakes, flood plains, and swamps.

If and when the Vietnamese, Laotians, and Khmers so decide, an aquaculture can be created there, as beneficent as that developed in the Yangtze and Yellow River lowlands: an aquaculture that dams and harnesses the floods, gives electric power to the entire region, provides a communication and transport network, irrigates agriculture, provides massive quantities of fish, and cements peace and cooperation among war-ravaged peoples.

A step in the right direction has been taken. A Committee for the Coordination of Investigations of the Lower Mekong Basin has been established, with a Secretariat in Bangkok, and far-reaching plans for the implementation of a large-scale river training program through the construction of dams has been formulated.

In England and Wales, fish farmers have formed a trade union to secure their rights.

In the United States, the American Farm Bureau is beginning to put its full weight on the side of aquaculture, in support of new legislation before Congress. The farm bill which Congress passed in 1977 makes the Department of Agriculture the main federal agency responsible for aquaculture development. The National Academy of Sciences completed a special study of the existing constraints on aquaculture development in the United States and has made a number of concrete recommendations for action.

At the international level, a Mexican study of the legal, social, and economic aspects of aquaculture, presented at the Kyoto conference in 1976, recommended the convening of a Conference on the Law of Aquaculture to advance the development of aquaculture for the purposes of generating employment, income, food, and recreation for millions of people. "Scientific management of aquatic resources," the study concluded, "must lead to the multiple uses of water, and among these aquaculture merits particular attention as it constitutes one of the best hopes for meeting realistically the uncertainties of the future. The United Nations should make every effort to give to the people of the world a legal instrument apt to conserve and enhance aquatic resources through the application and integration of aquaculture."

The United Nations forum that should provide a legal framework for the development of aquaculture is the ongoing Conference on the Law of the Sea. Strangely, that Conference completely ignores the subject. Tired and embittered, delegations haggle over shares in surpluses of natural resources that cease to exist with the collapse of one commercial fishery after another. Why do they not join forces, instead, to *create* new resources through aquaculture? These resources should be considered the *common heritage of mankind,* and should be managed for the benefit of all countries, especially the poorer ones. What is needed is a new law revolutionizing our concepts of ownership and sovereignty and transcending the legal bounds of a dying past.

15

A NEW LAW FOR AQUACULTURE

The new Law of the Sea should make four major provisions for aquaculture.

Proposed New Law/Part One

The first part of the law as it affects aquaculture must provide for water conservancy and management. Here the Conference on the Law of the Sea has indeed made a beginning, by legislating a number of new principles of conservation and pollution control. What it has not yet done, however, is to provide an institutional framework for the international enforcement of these new principles, to *manage the ocean environment,* and to interweave national and international legislation and management systems in a consistent and coherent way. Much of this should be passed down to the local level: the global law should be articulated and implemented in a series of regional conventions, such as the existing Convention for the Baltic Sea or the Barcelona Convention for the Mediterranean. It is then the turn of national legislatures to implement and complement the law. The road ahead is long and arduous. Without a clean aquatic environment there can be no aquaculture. Aquaculture itself may eventually make a major contribution to cleaning up the sea, through the creation of biological or bacteriological systems that destroy pollutants.

Proposed New Law/Part Two

The second part of the law must provide for the proper integration of aquaculture practices with other water uses and the definition of rights and duties of states and enterprises.

Freshwater and Brackish Water
With regard to freshwater and brackish-water culture, it is, obviously, national legislation that has to move, in the directions indicated. Given the interdependence of all parts of the aqua-ecosystem and the possible

222

effects of land-based activities on the marine environment, it is necessary, however, that even inland freshwater and brackish-water culture projects be planned in consultation with the competent international organizations. In the case of developing countries, these organizations are necessarily involved, because of the need for financial and technological assistance.

Oceans: (I) Areas Under National Jurisdiction

Oceanic culture, as we have seen, is characterized by two types of farming. One type is territorially limited. It includes the farming of sedentary organisms on the seabed such as mussels, oysters, and crustaceans; it includes cage culture, raft cultures, and embayments. This type of farming is done close to the shore, in areas under national jurisdiction. Although the new Law of the Sea extends the sovereign rights of coastal states over their nonliving and living natural resources and all related economic activities to 200 miles offshore, these rights cannot be extended automatically to cultured living resources. Cultured living resources are not natural resources. They are in a different category. And other states, for example, landlocked states, which engage in aquaculture in the coastal waters of a neighboring state, are not encroaching on the natural resources over which that state has sovereign rights. On the contrary, they are enriching these resources. Therefore there is no reason for excluding those states, as the framers of the emerging new Law of the Sea, aware only of traditional capture fishery, appear to be doing.

On the other hand, aquacultural activities, even within areas under coastal state jurisdiction, by introducing new species, hybridizing, or fertilizing, may alter the ecology of a whole region, or may alter it far beyond the limits of one state's jurisdiction. By exercising their own sovereign rights, states could destroy the sovereignty of other states, whose environment might be modified or altered without reference to their own sovereign decisions.

To safeguard national sovereignty, all states whose interests are affected by the decision should participate in the decision-making process.

International law for aquaculture in areas under national jurisdiction thus might take the following form:

The coastal state, in consultation with the competent regional or global international organization, with due regard to other uses of ocean space, such as navigation or the laying of cables and pipelines, designates certain areas as *aquaculture zones*. States which have no access to the sea (landlocked states) or which have otherwise no possibility of establishing aquaculture zones (some geographically disadvantaged states) have a right to participate in aquacultural activities in these zones, either at their own expense or, if the coastal state so wishes, in joint-venture with this state or one or more of its enterprises. Plans for aquacultural development, whether by the coastal state or by another state, must be approved by the competent international, regional, or global organization, of which that state, as well as all states that might be affected, obviously should be members.

Oceans: (II) Areas Under International Jurisdiction
The other type of seafarming is sea ranching, which extends in theory over whole oceans. Obviously, this type of farming cannot be fully developed so long as it is possible for one state to make substantial investments in hatching and nursing fingerlings for release in the open sea, only to have them caught by the capture fishermen of other nations. The new law of aquaculture must provide support through the competent international organization to all those states engaged in the hatching and nursing development. At the same time it must ensure a fair return in catches to all participating states. This could be done most economically by replacing the present competitive scramble, in which very little is caught by a far greater number of ships, manpower, and capital investment than would be needed by joint fishing enterprises, with the joint produce distributed according to need and degree of participation. This would undoubtedly leave a number of capture fishermen unemployed, but they could be retrained for aquaculture projects. Rational management requires—and has for some time required—a reduction in capture efforts. There will be more fish, with less effort, and at a lower price. This can be achieved only through international cooperation and through a new Law of the Sea.

Proposed New Law/Part Three

The third part of the new international Law of the Sea concerns protection against diseases of aquatic organisms and their transmission from country to country. The large-scale international trade of fingerlings and fish eggs—rainbow trout from Denmark, elvers from Western Europe to Japan, oyster eggs from the West Coast of the United States, and the transplanting of species to new marine environments—bass from the United States to the Soviet Union, Atlantic and Pacific salmon all over the world—entail the danger of spreading pests and plagues which may be hazardous for human beings, wipe out entire cultures and thus cause heavy economic losses, or even spread from cultured to wild animals in a given region.

A great deal of progress has been made in recent years in the study of freshwater fish diseases, but very little is as yet known about marine fish medicine. Five contagious fish diseases, in particular, have attracted wide attention: Infectious pancreatic necrosis, viral hemorrhagic septicemia, infectious dropsy, furunculosis, and whirling disease. About eel papilloma or eel branchia-nephritis or fish vibriosis, which may affect marine fish, far less is known.

The Japanese have built a remarkable fish hospital for the study of the prevention and treatment of fish diseases and for the study of fish breeding. The hospital, an interesting example of modern architecture, is located in Toba Bay, Mie Prefecture, on the Pacific Ocean. The staff cares for approximately 4,000 fish belonging to about 250 species, and sea turtles and porpoises. Staff members also make house calls to fish farms and home aquariums. Fish surgeons have developed a technique of anesthetizing fish in surgery tanks and performing the operation in the water. Other diseases can be cured with ingenious simplicity. Thus the popeye disease, afflicting many fishes in captivity, and caused by bacterial infection, can be cured by pouring some human eye lotion, obtainable in any drugstore, into the fishes' habitat.

National and international machinery has been set in motion to create a legal framework for the control of contagious fish diseases. The International Office of Epizootics in Paris has published an International Zoo-sanitary Code for Mammals, Birds, Fish, and Bees, as a guide to facilitate the work of veterinary services. It contains a model or pattern certificate for use in international trade in live fish and fish eggs. There is now a Permanent Commission for the Study of Fish Diseases which organizes international symposia and has elaborated an international prophylactic plan for fish diseases. And FAO has published a Comparative Study of Laws and Regulations Governing the International Traffic in Live Fish and Fish Eggs (1968). All this has led to the proposal for an International Convention for the Control of the Spread of Major Communicable Fish Diseases, which, after a series of technical conferences, is ready for signature. Such a Convention, providing, among other things, for inspection at the point of departure and the issuance of health certificates vouching for the absence of the major destructive diseases of aquatic organisms, must be part of the new Law of Aquaculture.

Proposed New Law/Part Four

Fourthly, aquaculture development must now be considered as an integral part of development in general. It is amazing that, until now, international resolutions, programs, or strategies for development of world food plans have simply ignored the potential contribution of marine resources and ocean management. It was somehow as though the ocean planners had been separated by an ocean from the rest of development planners. This may still have been understandable in the 1960s, when the "marine revolution," that is, the penetration of the Industrial Revolution into the oceans, was at its beginning. It was an anomaly in the 1970s when the multiple uses of ocean space and resources came to light at the great United Nations Conference on the Law of the Sea. In the 1980s, when marine products will constitute an even greater portion of the world GNP, it would be ludicrous to forget them. Aquaculture can be properly financed and fully developed only when it is properly integrated in world food plans and development strategies.

Integrated development of this sort, as pointed out by the 1975 World Food Conference, is a "very complex long-term process entailing changes in patterns of ownership, political power structures, social traditions and attitudes, the organization of economic activity and the institutional and administrative set-up of each society."

EPILOGUE

The growing importance of aquaculture is part of the ongoing profound transformation of humankind as it penetrates outer space and ocean space, atomic space, and some of the inner space of the mind. All these processes are concurrent.

The first conquest of ocean space, beginning at the time of the great explorers and continuing to our day, has been physical: the discovery of continents and islands on the surface of the planet, the mapping of the deep ocean floor. This first conquest of ocean space has had, and continues to have, a profound transformatory effect on the history of humankind.

The second conquest of ocean space is its transformation from wilderness to farm and mine—or even mining farm or biological mine.

Seafarm, to be sure, with its arms reaching deep into the land, will always remain part wilderness, untamed, where man must learn to live with nature, within nature, a part of nature, not over and against nature. Hoary fictions of his terrestrial mind, like sovereignty, ownership, boundaries, are ground and polished by the waves like rocks and pebbles. This transformation, again, will have a profound effect on the terrestrial history of humankind.

The second conquest of ocean space, its transformation from wilderness to farm, is a feat of even more imposing magnitude than the perilous voyages of the explorers. It is an undertaking that will keep us busy for the next several hundred years.

BIBLIOGRAPHY

Books

Aquaculture Planning in Africa. Report of the First Regional Workshop on Aquaculture Planning in Africa, Accra, Ghana, July 2-17, 1975. Rome: FAO, 1975.

Aquaculture Planning in Asia. Report on the Regional Workshop on Aquaculture Planning in Asia, Bangkok, Thailand, October 1-17, 1975. Rome: FAO, 1976.

Aquaculture Problems and Increasing the Bioproductivity of the World Ocean. Theses of the papers presented at the Sixth Japan-USSR Joint Symposium on Aquaculture, October 7-21, 1977, Moscow and Batumi. Moscow: VNIRO, 1977.

Bardach, John E.; Ryther, John H.; and McLarney, William O. *Aquaculture: The Farming and Husbandry of Freshwater and Marine Organisms.* New York-London-Sydney-Toronto: Wiley-Interscience, 1972.

Brown, E. Evan, *World Fish Farming, Cultivation and Economics.* Westport, Conn.: AVI Publishing, 1977.

Central Institute of Fisheries Technology. *Souvenir.* Cochin: 1976.

Committee for the Coordination of Investigations of the Lower Mekong Basin. *Feasibility Study on Pilot Fish Farm Projects in the Mekong Basin.* S. K. R. International Consultants Ltd., Israel, and Ministry for Foreign Affairs International Cooperation Division, State of Israel, August 1974.

Department of Fisheries, Ministry of Agriculture and Cooperatives, Thailand. *Shrimp Farming in Thailand.* 1973.

Department of Planning and Economic Development, State of Hawaii. *Aquaculture in Hawaii.* 1976.

Estes, Thomas S., ed. International Center for Marine Resource Development. San José, Costa Rica, January 13-17, 1975. *Proceedings of the Seminar Workshop on Artisan Fisheries Development and Aquaculture in Central America and Panama.* Kingston, Rhode Island: University of Rhode Island, 1976.

Fisheries and Marine Service, Ottawa. *Final Report: Visit of the Canadian Fisheries Mission to the Peoples Republic of China, November 22-December 9, 1974.* January 1976.

Freshwater Fisheries and Aquaculture in China. Rome: FAO, 1977.

Idyll, C. P. *The Sea Against Hunger.* New York: Thomas Y. Crowell, 1970.

Indian Council of Agricultural Research. Tuticorin, January 24, 1974. *Proceedings of the Group Discussion on Pearl Culture.* 1974. Cochin: Central Marine Fisheries Research Institute.

Instituto Nacional de Pesca: Información. *Catálogo de artes y métodos de pesca artesanales de la República Popular China.* 3 vols. Mexico: 1975.

Japan Marine Products Photo Materials Association. *Fisheries in Japan: Eel.* Supervision Tetuo Tomiyama, Takashi Hibiya. Tokyo: 1977.

———— *Fisheries in Japan: Prawns.* Supervision Tetuo Tomiyama. Tokyo: 1973.

———— *Fisheries in Japan: Sea Bream.* Supervision Tetuo Tomiyama. Tokyo: 1974.

Jhingran, V. G. *Fish and Fisheries of India.* Delhi: Hindustan Publishing Corporation, 1975.

Korringa, P. *Farming Marine Organisms Low in the Food Chain.* A Multidisciplinary Treatise. Development in Aquaculture and Fisheries Science, 1. Amsterdam-Oxford-New York: Elsevier Publishing Co., 1976.

———— *Farming the Cupped Oysters of the Genus Crassostrea.* A Multidisciplinary Treatise. Development in Aquaculture and Fisheries Science, 2. Amsterdam-Oxford-New York: Elsevier Publishing Co., 1976.

———— *Farming the Flat Oysters of the Genus Ostrea.* A Multidisciplinary Treatise. Development in Aquaculture and Fisheries Science, 3. Amsterdam-Oxford-New York: Elsevier Publishing Co., 1976.

———— *Farming Marine Fishes and Shrimps.* A Multidisciplinary Treatise. Development in Aquaculture and Fisheries Science, 4. Amsterdam-Oxford-New York: Elsevier Publishing Co., 1976.

Kurian, C. V., and Sebastian, V. O. *Prawns and Prawn Fisheries of India.* Delhi: Hindustan Publishing Co., 1976.

Madlener, Judith Cooper. *The Sea-vegetable Book. Foraging and Cooking Seaweeds.* New York: Clarkson & Potter, Inc., Publishers, 1977.

The Oceanic Institute. *Annual Report 1976.* Hawaii: 1977.

Tokai University. September 1976. *Proceedings of the Fifth Japan-Soviet Joint Symposium on Aquaculture.* Tokyo and Sapporo: March 1977.

Articles

"Applying Science and Technology to African Fisheries." *Standard Bank Review.* October 1974.

Borgese, Idyll, Cuivers, Mrozek, Policastro, Pryor, and Foott. "Mariculture" (a series of articles). *Oceans.* March/April 1979.

Boschi, E. E., and Scelzo, M. A. "El cultivo de cámaros comerciales Peneides en la Argentina y la posibilidad de su producción en mayor escala." FIR:AQ/Conf/76/E.40.

Caubere, J. L.; Lafon, R.; Rene, F.; and Sales, C. "Maturation et Ponte chez Penaeus Japonicus en captivité, essai de control de cette reproduction à Maguelone sur les côtes françaises." FIR:AQ/Conf/76/E. 49.

Coche, G. "A General Review of Cage Culture and Its Application in Africa." FIR:AQ/Conf/76/E. 72.

Deveau, L. E., and Castle, J. R. "The Industrial Development of Farmed Marine Algae: The Case-History of *Eucheuma* in the Philippines and U.S.A." FIR:AQ/Conf/76/E. 56. FAO Technical Conference on Aquaculture, Kyoto, Japan, May 26-June 2, 1976.

Dillin, O.W., Jr. "Aquaculture in the Southern United States." FIR:AQ/Conf/76/E. 26, February 1976.

Figueras, A. "Desarrollo Actual del cultivo del mejillón (*Mytilus edulis* L.) y posibilidades de expansión." FIR:AQ/Conf/76/R. 7.

Fujiya, M. "Coastal Culture of Yellowtail (*Seriola quinqueradiata*) and Red Seabream (*Sparus major*) in Japan." FIR:AQ/Conf/76/E. 53.

Furukawa, A. "The Aquaculture Industry in Japan—Its Present and Future." FIR:AQ/Conf/76/E. 71.

Glude, J. B. "Oyster Culture—a World Review." FIR:AQ/Conf. 76/R. 16.

Gopalakrishnan, V. "Status and Problems of Culture of Baitfish for the Skipjack Fishery in the Pacific Region." FIR:AQ/Conf/76/E. 2.

Hanamura, N. "Advances and Problems in Culture-based Fisheries in Japan." FIR:AQ/Conf/76/R. 18.

Hanson, Carol; Collier, Jeanne; Craven, John P.; and Sheets, George M. "Legal and Political Perspectives on Open Sea Mariculture."

Hirasawa, Y. and Walford, J. "The Economics of Kuruma-Ebi (Penaeus Japonicus) Shrimp Farming." FIR:AQ/Conf/76/R. 27.

Kamara, A. B.; McNeil, K.; and Quayle, D. B. "Tropical Mangrove Oyster Culture: Problems and Prospects." FIR:AQ/Conf/76/E. 58.

Koganezawa, A. "The Status of Pacific Oyster Culture in Japan." FIR:AQ/Conf/76/E. 69.

Korringa, P. "Economic Aspects of Mussel Farming." FIR:AQ/Conf/76/R. 3.

Laubier-Bonichon, A. and Laubier, L. "Reproduction controlée chez la crevette *Penaeus Japonicus*." FIR:AQ/Conf/76/E. 38.

Lindberg, J. M. "The Development of a Commercial Pacific Salmon Culture Business." FIR:AQ/Conf/76/E. 19.

McNeil, W. J. "Review of Transplantation and Artificial Recruitment of Anadromous Species." FIR:AQ/Conf/76/R. 24.

Mann, Roger. "Exotic Species in Aquaculture." *Oceanus.* Spring 1979.

Marderosian, Ara Der. "Marine Pharmaceuticals." *Journal of Pharmaceutical Sciences.* Vol. 58, No. 1, January 1969.

Meixner, R. "Culture of Pacific Oysters (*Crassostrea gigas*)." FIR:AQ/Conf/76/E. 28.

Michanek, Göran. *Seaweed Resources of the Oceans.* Rome: FAO Fisheries Technical Paper No. 138, 1975.

Milne, P. H. "Selection of Sites and Design of Cages, Fishpens and Net Enclosures for Aquaculture." FIR:AQ/Conf/76/R. 26.

Miura, Akio. "Present Practices and Future Potential in Seaweed Cultivation." *Ocean Yearbook.* Vol. II. University of Chicago Press, 1980.

Mizumoto, S. "Pearl Farming—a Review." FIR:AQ/Conf/76/R. 13.

Moller, D. "Recent Developments in Cage and Enclosure Aquaculture in Norway." FIR:AQ/Conf/76/R. 20.

Neal, R. A. "Penaeid Shrimp Culture Research at the National Marine Service Galveston Laboratory." FIR:AQ/Conf/76/E. 23.

Neish, I. C. "Culture of Algae and Seaweeds." FIR:AQ/Conf/76/R. 1, December 1975. Kyoto, Japan: FAO Technical Conference on Aquaculture, May 26-June 2, 1976.

Nikolic, M.; Bosch, A.; and Vázquez, B. "Las Experiencias en el Cultivo de Ostiones del Mangle (*Cassostrea rhizophorae*)." FIR:AQ/Conf/76/E. 52.

Pantulu, V. R. "Floating Cage Culture of Fish in the Lower Mekong Basin." FIR:AQ/Conf/76/E. 10. Kyoto, Japan: FAO Technical Conference on Aquaculture, May 26-June 2, 1976.

———. "Role of Aquaculture in Water-Resource Development—A Case Study of the Lower Mekong Basin Project." FIR:AQ/Conf/76/E. 20.

Patellani, Aldo. "Un Miracolo italiano: La Moltiplicazione dei Pesci." *Oggi* 33 (December 31, 1977): 63.

Pérez, D. Pérez and Suárez, M. Oliva. "Cultivo experimental de estadios larvales del camarón Blanco *Penaeus schmitti* y del Camarón acaramelado *Penaeus duorarum notialis* en laboratorio." FIR:AQ/Conf/76/E. 45.

Perrot, J. "Progrès dans les techniques d'élevage des crevettes et de la production de juveniles." FIR:AQ/Conf/76/R. 12, February 1976.

Pillay, T.V.R. "Progress of Aquaculture." *Ocean Yearbook.* Vol. I. University of Chicago Press, 1978.

Ryther, John H. "Aquaculture in China." *Oceanus.* Spring 1979.

Saito, Y. "Seaweed Aquaculture in the Northwest Pacific." FIR:AQ/Conf/76/R. 14, February 1976. Kyoto, Japan: FAO Technical Conference on Aquaculture, May 26-June 2, 1976.

Sandifer, P. I. and Smith, T. I. "Experimental Aquaculture of the Malaysian Prawn, *Macrobrachium Rosenbergii* (de Man) in Southern Carolina, U.S.A." FIR:AQ/Conf/76/E. 3, December 1975.

Standord, F. Bruce. "Seaweeds and Their Uses." Fishery Leaflet 469. Washington, D.C.: U.S. Department of the Interior, November 1958.

Stewart, J. E. and Castell, J. D. "Various Aspects of Culturing the American Lobster, *Homarus americanus*." FIR:AQ/Conf/76/E. 11.

Wilcox, Howard A. "The Ocean Food and Energy Project." No. 3, January 13, 1975, Public Affairs Office, Naval Undersea Center. A paper presented to the 141st Annual Meeting of the American Association for the Advancement of Science, January 29, 1975, in New York City.

INDEX

231

234

ACKNOWLEDGMENTS

The generous assistance extended to me while I was working on this book—by institutions, governmental and international officials, scholars, and seafarmers in all parts of the world—is in itself proof, in a way, that the living resources of Planet Ocean are indeed the Common Heritage of Mankind.

My special thanks go to John Bardach and Peter Korringa, on whose own work I have freely drawn in this presentation. Professor Bardach organized my research in Hawaii. Professor Korringa put at my disposal his unique collection of slides on aquaculture scenes and operations.

I am equally grateful to R.V. K. Pillay of FAO who helped to put together my Southeast Asian study trip. Dr. A. S. Bogdanov of VNIRO organized my research in the Soviet Union, Professor Igor A. Burtsoff helped with his invaluable experience with sturgeon breeding, and Michael Gaba of VNIRO was my resourceful travel companion from the Baltic to the Black Sea.

The late N. K. Panikkar was my host and mentor during my work in India, and my thanks go to his family.

The following have been particularly helpful during the Southeast Asian study trip, and I want to thank them here, together with many others, whom I fondly remember but cannot enumerate here:

G. P. Kumara S. Achari, Pilot Project on Pearl Culture, Department of Fisheries, Vizhingam, Trivandrum, Kerala.

Chertchai Amatyakul, Director General of Fisheries, Bangkok, Thailand.

José Buenaventura, Igang Seafarming Station, Philippines.

Chaliang Chaitiamvong, M. S. Freshwasher Fisheries Division, Department of Fisheries, Bangkok, Thailand.

Kulsak Chotiyaputta, Brackishwater Fisheries Division, Ministry of Agriculture and Cooperatives, Bangkok, Thailand.

Mahmood Ebrahim, Director, Indo-Marine Agencies, Cochin, Kerala.

José J. Estrella, Jr., Bounty Agro Fisheries Corporation, Philippines.

Dr. Sergio S. Felix, Chairman, Laguna Lake Fishpen Committee, Philippines.

Dr. A. N. Ghosh, Project Director, State Fisheries Development Corporation, West Bengal, India.

Dr. Felix Gonzales, Director, Bureau of Fisheries, Manila, Philippines.

Eliseo Griño, SEAFDEC, Iloilo, Philippines.

Dr. Kurien Jacobs, Department of Science, Cochin.

Abdul Kahar, Dinas Perikanan Properties, Semarang, Indonesia.

Professor N. B. Nair, Dean, Faculty of Sciences, University of Kerala, Trivandrum.

Dr. Nonaka, Shizuoka Prefecture Fisheries Research Station, Japan.

Dr. Herminio Rabanal, South China Sea Fisheries, Manila, Philippines.

M. K. Rangarajan, Madras Research Centre of the Central Marine Fisheries Research Institute, India.

Dr. Ariya Sidthimunka, Director, Inland Fisheries Division, Department of Fisheries, Thailand.

M. Soehardi, Director General of Fisheries, Jakarta, Indonesia.

M. Tulasilingam, Central Marine Fisheries, Madras, India.

Professor Jun Ui, University of Tokyo, Japan.

Professor N. K. Velankar, Department of Industrial Fisheries, University of Cochin, Kerala, India.

Swasdi Wongsommuk, Head, Aquaculture Division, Songkhla Fisheries Station, Thailand.

Dr. Isao Yano, National Pearl Research Laboratory, Kashikoyima, Mie, Japan.

Wilfredo G. Yap, Aquaculture Department, Southeast Asian Fisheries Division, Philippines.

Robert Hart of the Fisheries Department of Canada has given me full access to the written and pictorial records of his experience in the People's Republic of China, and I thank him for his generosity.

In the chapter on China I have also drawn freely on the Report of the FAO Mission to China.

I want to thank Ruth Boulez, who helped with the research on seaweeds and contributed her family experience with crayfish farming; the photographer Robert Ketchum, who shared much of the excitement of the Southeast Asian trip and, last but not least, my assistant Jean de Muller who helped with everything.

I am, as always, grateful to the editorial staff of Harry N. Abrams, Inc., to Michael George and especially to Margaret Kaplan, whose expertise and spirit of cooperation have been quite invaluable. The layout of the book is due, once more, to the special genius of Nai Y. Chang.

Without Dr. Fritz Landshoff, Executive Vice President of Harry N. Abrams, Inc., the book would never have been written.

Elisabeth Mann Borgese

The publishers gratefully acknowledge the generous assistance, through Michael Phillips, of Nikon, Inc.; also the help of the Vivitar Corporation.